The *Struan*

The *Struan* Name Board

The Norwegian wooden full-rigger *Struan* was originally a British vessel built by John Fraser at Saint John, New Brunswick in 1877. Rudderless, dismasted and waterlogged by a series of violent storms, and kept afloat only by the million-odd feet of timber in its hold, the ship was abandoned by its crew off the southern Oregon coast on December 15, 1890. The *Tamar E. Marshall*, Captain Andrew F. Kenney, of Digby, Nova Scotia, took all the *Struan* crew members aboard and transferred them to the tug *Vigilant* just outside San Francisco Bay about three weeks later. On Christmas day the derelict *Struan* stranded on the beach a mile south of Cape Lookout and broke up within four hours, scattering its cargo for 20 miles along the coast and providing a bonanza of building materials for early settlers and homesteaders in the area.

Early in January 1891, one of the *Struan*'s two name boards was picked up on the shore by a member of the Chamberlain family of homesteaders. To dry the board slowly, and avoid checking or cracking, it was stored in the unheated family barn.

When the Chamberlains moved away from the homestead in 1904, to a more urban location where the grandsons of the original settlers could obtain a better and more advanced education than was then available in the small rural local community, the name board was either forgotten or considered not worth taking.

The Portland Area Council of the Boy Scouts of America purchased the homestead property from the Chamberlain heirs in 1926. By that time, the house itself had burned to the ground, but the barn, though in poor condition, still stood. The *Struan* name board was rescued when the barn was dismantled so that some of the better lumber used in its construction could be utilized in early buildings at Camp Meriwether.

Old timers at the camp recall that the carved-out lettering and ornamental filigree work on the name board have been lovingly repainted at least twice by scout craftsmen in the past 60 years.

Some viewers, observing that the length of the *Struan* board could easily have accommodated an eight or even ten-letter vessel name, have wondered why such a long work piece was selected. The answer may lie in the fact that in the glory days of wooden shipbuilding in New Brunswick and Nova Scotia, apprentice carvers were kept busy preparing ornamental blanks for identification of the many ships being launched each year. Employers were not anxious for their artisans in training to spend time and effort on shorter boards that would be suitable only for the minority of vessels with very short names.

Shipowners or shipbuilders questioning the use of the longer boards for relatively short names would have been reminded that a substantial membership of the maritime fraternity subscribed to the theory that vessel names could be more readily made out through telescopes or spyglasses in chance encounters at sea (as in sightings or speakings) when the letters in the name board were disproportionately separated and could be individually recognized.

The *Struan*: From Saint John to Sandlake

For all the Scouts
in Troop 852
Whit Crichton
april 24, 2005

NIMBUS
PUBLISHING

Nimbus Publishing Limited
PO Box 9301, Station A
Halifax, NS B3K 5N5
(902) 455-4286

Page layout: Terri Strickland
Printed and bound in Canada

Canadian Cataloguing in Publication Data
Crichton, Whitcomb
The *Struan*
ISBN: 1-55109-287-5
Includes bibliographical references and index.

1. Struan (Ship). I. Title
VM395.577C74 1999 387.2'24 C99-950142-9

Nimbus Publishing acknowledges the financial support of the Canada Council and the Department of Canadian Heritage.

Table of Contents

Acknowledgments

The cooperation I have received in connection with preparation of the *Struan* history has been outstanding. Almost every source I consulted during the search was able to provide at least some small tidbit of information to contribute to the ever-developing picture of the *Struan*'s years afloat.

A few individuals deserve special mention:

Harry Allen of Donaghadee, County Down, for his dogged perseverance in a search for, and eventual discovery of, personnel records and other documents in possession of the Public Record Office of Northern Ireland (PRONI) in Belfast.

Suely Rodrigues Andrade, Rio de Janeiro, for her detailed translations from Portuguese to English of local newspaper accounts of the *Struan*'s two visits to Rio de Janeiro.

Professor Thomas Birnie, Portland State University, for preparation of requests in their language to German authorities in Hamburg, and for English translations to answers from that location; for translations to English of records, diaries, letters and documents discovered in Norway during a three-year search.

Elizabeth Anne Cherednik, then on staff at the Moore Branch Library, Tacoma, Washington, for her continuously cheerful and enthusiastic help in assembling older maritime volumes in support of my brother Don Crichton's invaluable assistance relating to the technological background of the *Struan*'s history.

Kare Olsen of the Riksarkivet staff in Oslo, for assembling a photocopied collection of *Struan* documentation (which was made available to organizers here in time for the *Struan* Shipwreck Centennial in 1990) and for arranging for the author to be admitted in 1991 to the three-floors-below-ground-level vault in which the original *Struan* records and other vessel histories are preserved.

Marshall Turner of Portland, Oregon, for composing initial inquiry letters in Spanish for Chilean sources.

Many libraries, museums, universities and other institutions throughout the world were consulted, mostly on personal visits, more than a few on multiple occasions. In some cases, written or e-mail answers to specific questions were provided upon inquiry to a particular institution.

To the best of my recollection, the following is a list of the institutions visited or consulted:

Alaska State Library, Juneau
Astoria Public Library, Astoria, Oregon
Bandon Public Library, Bandon, Oregon
Bibliotek, Haugesund, Norway
Boston Public Library, Massachusetts
British Newspaper Library at Colindale, North of London
City Council and Library, Newcastle, New South Wales, Australia
District Record Office, Stavanger, Norway
Eureka Public Library, Eureka, California
Florida State Library, Tallahassee, Florida
Glasgow Public Library, Scotland
Guildhall Library, Aldermanbury, London
In-house Library, Pensacola Grand Hotel, Pensacola, Florida
Kirn Memorial Library, Norfolk, Virginia
Library of Florida State University, Tallahassee, Florida
Linen Hall Library, Belfast, Northern Ireland
Main Library, Halifax, Nova Scotia
Mariners' Museum and Maritime Library, Newport News, Virginia
Maritime Museum Prins Hendrik, Rotterdam
Maritime History Archives, Memorial University of Newfoundland, St. John's
Maritime Museum and Library, Bergen, Norway
Maritime Museum of the Atlantic, Halifax, Nova Scotia
Market Square Library, Saint John, New Brunswick
Multnomah County Library and Inter-Library Loan Facility, Portland, Oregon
Museum at Skudeneshavn, Norway
National Maritime Museum and J. Porter Shaw Library, Fort Mason, San Francisco, California
National Maritime Library and Museum, Greenwich, England
National Museum of Wales, Cardiff
New Orleans Public Library, Louisiana
Pace Library at the University of West Florida, Pensacola

Pensacola Public Library, Florida
Public Archives of Nova Scotia, Halifax
Public Library of Cardiff, Wales
Public Library, Coquille, Oregon
Public Record Office of Northern Ireland (PRONI), Belfast, Northern Ireland
Public Records Office, Chancery Lane, London
Public Records Office, Kew Richmond Surrey, England
Rijksmuseum Nederlands Scheepvaart Museum, Amsterdam
Royal Norwegian Archives (Riksarkivet) Oslo, Norway
San Francisco Public Library
Tacoma Public Library, Tacoma, Washington
Tallahassee Public Library, Florida
Tillamook Public Library, Tillamook, Oregon
University of Alaska, Southeast, Library at Juneau
University of New Brunswick Libraries at Fredericton and Saint John
Upsala University Computer Center, Sweden
Vancouver Public Library, Vancouver, British Columbia

Preface

One of the most difficult problems involved in gathering information for any historical record is separating fact from fiction. Newspaper articles at the time of an accident or other incident one hundred years ago are sometimes found to have been coloured by the imagination of the reporters or those from whom they obtained their information.

Some news accounts concerning loss of the *Struan* were suspect by their very content. One story in a Pacific Coast daily a few days after the stranding reported that the ship "was owned by a poor family in Haugesund, and in the loss they have lost everything they had in the world," is obviously suspect, since the value of such a ship and its cargo would have exceeded the lifetime earnings of an average family at that time and place.

A usually respected marine history book published soon after the wreck (and in later editions) contributed to the confusion and uncertainty regarding the *Struan*'s loss by carelessly misspelling the ship's name as *Straun* and mistakenly identifying the rescue vessel by its first name, ascribing the second name to its captain.

Contrary to the erroneous statements of newspaper reports and published authors of the day, all 18 members of the crew of the *Struan* were rescued at sea from their drifting and waterlogged vessel, before it was cast up on the Oregon shore, by the British bark *Tamar E. Marshall* of Digby, Nova Scotia, and were safely landed in San Francisco on January 2, 1891.

One newswriter included in his story a statement that "the Captain and crew of 26 abandoned the vessel and have never been heard from." That information was still being accepted as fact fifty years or more after the entire crew had reached safety at San Francisco and the captain, two officers, the deck boy, and a crewmember who deserted in South America, had returned to Norway to pursue their seafaring careers.

Newspapers did not give as much attention to the arrival in San Francisco of the *Struan* crew on January 2, 1891, as they had accorded the story of the shipwreck itself. Because telegraphic dispatches to other Pacific Coast big-city dailies failed to report the crew's safe arrival, Tillamook County residents near the wreck site, who depended for their out-of-the-county news on mailed copies of the Portland *Oregonian*, had no way of knowing that all members of the ship's company had survived the *Struan*'s loss.

One reporter's careful recollection of conversations with the shipwrecked sailors on their being set ashore, proved to be a starting point for the thread that would unravel more of the *Struan* story. He mentioned in his article that some crew members had expressed regret that the third mate had been seriously injured in a fall from the rigging while setting a distress signal upon the approach of the rescue vessel "since he had been on board only about three weeks." That little scrap of information enabled us to dig out records in the Royal Norwegian Archives (Riksarkivet) in Oslo verifying that the third mate did in fact sign on at Port Townsend, Washington, just before the *Struan* left that harbor on its final voyage.

Generally speaking, the smaller newspapers managed to print more accurate stories of the incident than did the larger ones. While some of the Pacific Coast big-city dailies published telegraphic dispatches from San Francisco incorporating and repeating whatever errors were present in the original version, a small Tillamook County weekly printed a letter from an eyewitness to the stranding. That article became the basis for what was factually known of the wreck for over fifty years.

It is understandable that occasionally newspaper employees not well acquainted with ships and shipping were assigned to write articles that required the help of someone more knowledgeable in seafaring terms and practices.

The intricacies of the rigging on sailing vessels one hundred years ago were probably confusing to people not familiar with the differences involving various types of vessels.

So it is easy to understand why, in some dispatches, full-rigged ships were sometimes identified as barks, schooners, sloops or whatnot, without regard to what type of vessel they actually were. This practice would be closely akin to labeling a 747 jet aircraft today as an L-1011 or a DC-10.

As well, failure to become familiar with the long-established system for determining and identifying latitude and longitude has led many writers to quote locations for newsworthy maritime events that were hundreds or perhaps even thousands of miles from the actual place of occurrence, or in some cases in the middle of a desert or on a mountain top.

Another criticism that can be made of a reporters, editors or other writers, involves their careless or sometimes intentional disregard for the actual facts involved in an event.

For instance, this writer recently read a story in a seaport-located British daily in which the details of the rescue of the crew from a dismasted and waterlogged sailing vessel were completely changed before publication. This may have been done to cover up the fact that the temporary or relief captain of a British-registered vessel had almost precipitated a minor international incident by petitioning the wrecked vessel's government to pay for food provided to the shipwrecked mariners during the two-week-plus period they were aboard his vessel before they could be put ashore.

When the *Struan* deck boy's recollection of the vessel's last trip was discovered—it had been written for a Norwegian Shipping Jubilee in 1935—we assumed that a gold mine of authentic detail was finally available. When the document was translated from the original Norwegian, however, questions were raised about its accuracy because the facts didn't jibe with certain Lloyd's Register accounts of events during the voyage, or with statements made immediately after the vessel's loss by the captain in his report to the owners regarding the wreck.

Later research revealed that the intervening 45 years between the wreck and the writing of the story had clouded deck boy Madsen's memory and confused some details to the point that his treatise could not be relied upon as an accurate description of the *Struan*'s day-to-day activities.

An English translation of Madsen's account was published in a Sunday Supplement to the Portland *Oregonian* September 5, 1954, and included a photograph of a painting of the ship *Struan* in the Haugesund museum. That original article has been plagiarized, added to and changed by a 40-year succession of so-called "feature writers." In some of these copy-cat presentations, for instance, ports of call have been added, facts altered, embellishments made, and even the place where the vessel had been constructed identified incorrectly. In one instance, the only picture of the ship available at the time was "doctored" to accompany the story of a "ghost ship"—a disgrace to historical reporting.

Most writers on the subject of the *Struan* have copied earlier accounts without bothering to do any investigative work on their own, even to the extent of blindly repeating the incorrect spelling of the vessel's name.

A pamphlet on shipwrecks currently being sold along the Pacific Coast correctly identifies the wreck site, but illustrates the derelict *Struan* stranded on the beach with all masts and yards intact as though standing ready for an inspection by an Admiral of the Fleet at a London dockyard.

Experience has shown that even such usually respected sources as encyclopedias are not always correct. In its 1952 edition, one well-known encyclopedic publisher illustrated an article on international code flags (for ship-to-ship and ship-to-shore use) showing the letters C, D, E, F and G as pennant-shaped (triangular) flags, although these had been superseded by new rectangular flags (to match the others in the alphabet) in a general revision of the International Code of Signals in 1931.

One published account of the *Struan*'s loss estimated the value of the ship and its cargo at $200,000, probably more than three times the vessel's cost a year earlier ($45,000), and the timbers loaded at Port Discovery (at the most $15,000) added together.

In this book we have made a conscientious and determined effort to produce a treatise that is historically correct to the smallest detail. If information we have discovered in various places throughout the world cannot be verified by official records, or by two or more independent other sources, it is not reported as fact, but qualified by the phrase "it is believed that" or some similar wording so that the reader may judge for himself or herself whether that evidence merits acceptance as fact, as a reasonable probability, or as a mere possibility.

Chapter I

The *Clythia* Marbles

One of the first volumes I examined in an early search for information concerning the *Struan* was a book by Esther Clark Wright of Wolfville, Nova Scotia, a prolific writer about the early settlements in the Canadian provinces of New Brunswick and Nova Scotia. Wright was a Doctor of Philosophy at Acadia University, who delighted in bringing to print information about the daily lives of the early pioneers in her part of Canada.

Many of her 15 published works dealt with subjects relating to the people, places, hardships and disappointments of early settlers in the two provinces across the Bay of Fundy from each other.

In 1976 she published a book tracing the history of wooden shipbuilding in and near Saint John, New Brunswick. In her discussion of John Fraser's yard at Courtenay Bay in Saint John,[1] she listed the *Struan* as his second ship and noted that "as the *Clythia*, it had been abandoned in the North Pacific in 1890."

I couldn't understand the basis for that report when I read it, since for almost sixty years I had been familiar with the *Struan* name board which had been picked up on the beach by a homesteader after the Christmas day stranding in 1890, and displayed at one place or another on the premises of Camp Meriwether (established in 1927), most recently on the wall of the dining hall.

No record of a vessel of *any* type with the name *Clythia* could be found in either Lloyds' or Bureau Veritas ship lists for the time between the *Struan*'s launching in 1877 and her stranding in 1890. It therefore became apparent that the reference to loss of the *Struan* under the name *Clythia* was not factual.

The challenge remained, however, to discover how the *Clythia*

---

1 Esther Clark Wright, *Saint John Ships and Their Builders*, published by the author, 1976, p.135.

name had crept into Wright's book and, if possible, to ascertain the facts about the vessel if one had ever existed. Records published annually by the New Brunswick government regarding name changes for vessels were searched to no avail.

When the staff of a Nova Scotia museum suggested that the name change might have been made coincident with the sale of the ship to Norwegian interests in the Fall of 1889, we solicited cooperation of the Royal Norwegian archives in Oslo. But there was no evidence there connecting the name *Clythia* with the *Struan*. The *Struan* had been purchased directly by Haugesund shipowners from British interests, and all transfer papers and Norwegian registry documents retained the *Struan* name designation.

An attempt to contact Dr. Wright by telephone in 1988 was unsuccessful. Subsequent correspondence with Acadia University revealed that she was then over ninety years old, in poor health, and not able to answer correspondence.

During my visit to New Brunswick and Nova Scotia in the summer of 1991, I learned that Dr. Wright had passed away on June 20, 1990, leaving some fifty boxes of notes and records to the Acadia University archives. I also discovered that her archival material was not catalogued or cross-referenced in such a manner that allowed the *Clythia* thread to be located.

Just as it appeared that the *Clythia* connection would never be satisfactorily explained, the mystery solved itself.

To refresh my memory of United States salvage law and to form some estimate of what the salvage award might have been if Captain Johnson of the tug *Scotia* had been able to tow the dismasted derelict *Struan* to a port of safety,[2] I picked up a copy of the leading treatise on United States salvage law.[3]

Captain P. A. Johnson of the steamer *Scotia* had boarded the derelict *Struan* about 3:00 p.m. on December 24, 1890 in latitude 45° 6' N and longitude 125° 10' W. When he discovered that the *Struan* was abandoned, he attempted to tow the ship to a port of safety, but found her so heavy with water that his vessel could hardly move her. After the hawser parted twice and a strong southeast gale and accompanying rough sea interrupted his efforts, Johnson realized that his prize was in danger of proving the *Scotia's* destruction. Rather than risk his own vessel, he cast off the tow line and proceeded on his

---

2 The (Portland) *Oregonian*, December 29, 1890.

3 Martin J. Norris, *The Law of Salvage*, Baker, Voorhis & Co., 1958. (1970 cumulative supplement, p. 104).

way. (If Captain Johnson's location of his encounter with the *Struan* is correct, the derelict drifted approximately 30 miles in a northeasterly direction prior to stranding on the beach a mile south of Cape Lookout before dawn the next day).

On the very page to which I opened the book was a discussion of the federal court case concerning the salvage in 1960 of part of the cargo of the Norwegian barkentine[4] *Clythia*, wrecked on the *east* coast of the United States.

Newspaper reports in the Norfolk, Virginia area at the time of the *Clythia*'s loss favor classification of the stranded vessel as a *bark.* A sketch of the wreck, made from an old photograph, lends credence to the *bark* theory, since the two forward masts appear taller and more suitable for square-rigged sails. The mizzen mast seems shorter, probably designed for the fore-and-aft sail carried in that position on a bark. The number of men in the crew (17) also indicates the likelihood the *Clythia* was a *bark*, since 17 or 18 crewmembers were usually carried by such vessels at the time. A *barkentine*, especially when operated by frugal Norwegian owners, could probably have been safely manned with two or three fewer crewmen because there would have been only one mast carrying square-rigged sails.

Had the *Clythia* been rigged as a barkentine, it would have carried square sails only on the foremast, with fore-and-aft rigging on the main and mizzen masts.

It is possible that the word "bark" was used incorrectly as an abbreviation for "barkentine," or more unlikely, that the vessel had been re-rigged.

The *Clythia*, inbound for Baltimore, Maryland, went aground January 22, 1894 (fully three years after the *Struan*'s loss) about 250 yards off the shore of Princess Anne County, Virginia near the Virginia border with North Carolina.

In thick fog with a heavy surf running, the vessel ran bow-first onto Pebble Shoal very early in the morning about midway between high tides. Its predicament was soon discovered by the north patrol from Wash Woods, North Carolina lifesaving station. When that keeper decided the sea was too rough for a through-the-surf boat launching, the crew turned out at once with its apparatus cart containing breeches buoy equipment and rigging. By the time they had completed their struggle along the shoreline to a point opposite the

---

[4] Legal records pertaining to the salvage of part of the cargo of the *Clythia* in 1960 consistently refer to the vessel as a *barkentine*. The 1891-92 edition of Lloyds' Register of Shipping classifies the *Clythia* as an iron *bark* of 1,140 net register tons.

wreck, the incoming tide had not only floated the stranded vessel over Pebble Shoal closer to shore, still bow-on to the beach, but was so far in that the would-be rescuers found themselves working in ankle-deep water while they tried to prepare a stable base upon which to set up their Lyle gun. By the time another lifesaving crew from False Cape station arrived, a hastily erected platform of driftwood and sand had been built, on which the gun was quickly mounted.

Fortunately, the first shot was good, and the breeches buoy was in service within a very short time. The first man ashore was the captain, who wanted to know his location. Like dozens of other shipmasters before him and after him, he had missed the entrance to Chesapeake Bay, and thought he was off Cape Henry, miles to the north.

Several witnesses who were present[5] have detailed the safe landing of the entire crew, 17 persons, of whom eight were identified as Italians, seven as Swedish, and one black man whose nationality was not recorded.[6] The captain must be presumed to have been Norwegian, since his vessel was not only registered in Norway, but in Norway's capital city, Christiania (now Oslo).

News of the stranding of the *Clythia* found its way to the front page of the Norfolk, Virginia Public Ledger on Monday afternoon, January 22, 1894, but only as one item in a column of miscellaneous local news. "Vessel Ashore" read the brief entry:

> The Norwegian barkentine *Clythia*, from Genoa for Baltimore, went ashore last night at Wash Woods, N.C. Four of the crew have been landed by the Lifesaving Service and the rest will soon be brought to shore. The Merritt Wrecking Company's steamer *William Coley* will leave for her assistance this evening.

The next afternoon an even shorter report advised readers that "the Merritt Company's steamer *I.J Merritt* is at work on the bark *Clythia* at Wash Woods. All the crew have been saved. The bark is loaded with marble and furniture."

---

[5] D. Gregory Claiborne Butts, *From Saddle to City, By Buggy, Boat and Railway*, Virginia Conference of the Methodist Episcopal Church, South, Hilton Village, Virginia, 1922, pp. 194-195.

[6] At that time, Sweden and Norway did not have separate consulates. Citizens of both countries were served by a combined Swedish and Norwegian consular service. It is likely that witnesses at the site of the stranding assumed that the members of the crew and the captain of the stricken vessel were Swedish because they asked to be put in touch with Swedish consular authorities.

Apparently a shipwrecked vessel was such a common occurrence near a large seaport in those days that the news of such an event evoked no more public interest than the next paragraph in the same column, relating that "the fine new carriage of Mr. David Dunlap was badly damaged by the running away of the horses on Washington Street this afternoon."

The official published records of the U.S. Lifesaving Service for 1894 enumerate the additional help provided by the Wash Woods station. Rescued crew members, insurance representatives and salvage vessel operators were fed and sheltered at the station until February 6, when the *Clythia's* captain finally gave up his bark for lost.

On a subsequent high tide, the *Clythia* broached broadside to the shoreline and the hull was apparently punctured, either by shifting of the cargo or popping of some hull rivets. From that time, the vessel is not believed to have moved an inch, except perhaps to dig itself deeper into the sand as the years passed.

Two salvage partners, E.H. Wiggins and W.E. Moulton, used explosives in 1960 to blast loose the iron deck beams and open up the badly rusted hull to make possible the recovery of the 12-ton marble blocks by lifting them with a railroad crane attached to their salvage vessel, the converted infantry landing craft *Big Wig*. They later raised to the surface and safely transported to shore some 123 tons of the cargo of Italian statuary marble which had been loaded in Genoa. A second visit to the *Clythia* wreck site almost resulted in loss of the *Big Wig* itself when the vessel was blown over the old wreck by tropical storm *Brenda*, had its hull punctured, and was beached when its anchor failed to hold. Captain Wiggins, with his partner and crew, was able to refloat the temporarily repaired salvage craft a few days later by securing a cable to the further-offshore wreck of the *Clythia* and pulling the *Big Wig* off the beach at high tide with its own winch.

Almost immediately upon bringing the salvaged marble ashore, the salvors were challenged by other divers who had previously obtained what purported to be an exclusive salvage license issued by the Virginia Commissioner of Wrecks *prior* to Wiggins' and Moulton's successful effort at the wreck site; the challengers, however, had made no effort to exercise that license.

The recovered marble was impounded by court order, and held until the verdict was rendered.

Judge Walter E. Hoffman of the U.S. District Court for the Eastern District of Virginia, Norfolk Division, ruled on September 12, 1960 that those persons holding an exclusive salvage license from the Virginia Commissioner of Wrecks, though they had dived several times to inspect and/or examine the cargo in the wreck, "had not performed a single affirmative act of actual salvage, and so had not established possession of the ship and were not entitled to salvage rights in the 123 tons of marble raised from the ship by other persons."

He found further that the *Clythia*, having remained in one spot for 66 years without any apparent attempt by its owners or their representatives to reclaim the vessel or its cargo, had been abandoned under the doctrine of *Animus Revertendi*—the owner has no intention of returning—and was therefore subject to appropriation by anyone.[7]

The authority of the Virginia Commissioner of Wrecks to issue an exclusive salvage license involving a wrecked vessel regarding which neither he nor his predecessors in office had ever taken any action in more than 65 years, was considered by the judge to be no greater than that of any other individual. Further, it was held, the Commissioner of Wrecks, having never appropriated or taken possession of the *Clythia* or her cargo, was not entitled to the commission specified in 62-165 of the Code of Virginia 1950 (five percent of salvaged material to the value of $10,000 and two percent thereafter).

The *Struan* was a wooden full-rigger loaded with timber, and quickly broke up in the surf when stranded, scattering its cargo and sections of its hull for twenty miles along the *west* coast of the United States and covering the beach in the immediate area of the wreck to a depth of six feet with ship's supplies and debris of every imaginable description. The *Clythia* was an iron bark which had been renamed just prior to its loss on the U.S. *east* coast.[8] The latter vessel was carrying some 1,100 tons of Italian marble, and sank like a rock when its hull was pierced after grounding.

References to a glistening white marble figurehead on the *Clythia* were made in several newspaper accounts immediately after the wreck and by a later writer in an account of east coast shipwrecks. In two 1992 interviews with Captain Wiggins, I learned that he had no first-hand information regarding the figurehead, though he reported that it was generally believed by oldtimers in the area that the ordinary wooden figurehead had been in the possession of False Cape Lodge soon after the wreck. Having been left out in the weather for many years without care or preservation, it was considered beyond any hope

---

[7] *E. H. Wiggins and W. E. Moulton, Libellants v. 1100 Tons, More or Less of Italian Marble*, Respondents; No. 8115, 186 Fed Supp 452, 457 (1960).

[8] The *Clythia* was built in Scotland in 1860 as the *Clyde*, and was operated until 1891 under that nomenclature, at which time the name was changed to *Clythia*. The hull plating was thicker even than that required under 1891 rules. It last received a special survey (inspection) at Liverpool in May 1892.

of reconditioning when it presumably came into the possession of a museum in the Norfolk area.

Captain Wiggins also contributed an interesting sidelight on construction of the *Clythia*. Although built in Scotland in 1860 as an iron vessel, each alternate rib in the hull was of wood, as if the design and construction had been handled by shipwrights whose experience was with wooden vessels, and the builders were not certain that iron was a suitable substitute. In the late 19th and early 20th centuries, when the transition from wood to iron was taking place, vessels using elements of the two methods in the same hull, were officially classified as composite vessels.

Another variation from present-day practice, Wiggins noted, was that the bark had only a wooden plank weather deck, laid directly across the iron deck beams.

Any historic interest in the figurehead might be increased by the fact that the *Clythia* sailed under the name *Clyde* for more than thirty years, presumably with a male figure on the stem-head. Unless a new figurehead was installed when the vessel was renamed *Clythia* in 1891, the Norwegian captain sailed his female-named vessel around the world possibly carrying a likeness of the Earl of Clyde or some other male figure at the bow.

In spite of being awarded title to the recovered marble by the court, Captain Wiggins and his partner, W. E. Moulton, made no profit on their salvage operation. Because sea worms had burrowed three or four inches into all the surfaces of the marble blocks, the material was considered unsuitable for any use, and no sale for even a portion of the recovered marble was ever made. Today the fruits of the salvage operation, moved two or three times since it was first unloaded onto the Pennsylvania Railroad docks in Norfolk, lie in a heap on the grounds of the discontinued Roanoke River lighthouse, which has been Captain Wiggins' home since he purchased the property on abandonment of the light by the government. Had Wiggins and Moulton researched the project more thoroughly before they began their salvage operation, they might have learned that a book published in 1952 carried the information that a diver *in 1925* had discovered the marble on the *Clythia* wreck worm-eaten and therefore worthless![9]

Even the satisfaction of having a lasting legal precedent established as a result of their litigation over the recovered marble cargo—that 66

---

9 David Stock, *Graveyard of the Atlantic – Shipwrecks of the North Carolina Coast*, University of North Carolina Press, 1952, p. 148.

years was a sufficient time for a court-ordered determination that the owner will not return to the ship or its cargo (*animus revertendi*)—was denied to Wiggins and Moulton. In another complicated maritime case involving the side-wheel wooden passenger steamer *Central America*, which began its progression through the U.S. Federal Court system in 1987, the Fourth Circuit Court of Appeals reversed the trial court verdict, effectively ruling that even 130 years in an unknown location a mile-and-a-half under the sea did not, in itself, prove abandonment. It was ordered that the *Central America* case proceed under the law of *salvage*, whereas the *Clythia* case had been judged under the law of finds.

Dr. Wright's intention in preparing the text for her *Saint John Ships and Their Builders* was obviously to gather material relating only to vessels built in Saint John and its environs. How an erroneous reference to the *Clythia* crept into her information stream remains a mystery although we can speculate that her source was unreliable, or that a researcher may have made a mistake while assembling material.

Fortunately, correct information about both vessels is now a matter of public record, and those documents prove conclusively that the only link between these two vessels was their Norwegian registrations, the *Struan* in Haugesund and the *Clythia* in Christiania (now Oslo). The ownership of each was independent of the other, though both were Norwegian-flagged at the time of their respective losses.

Chapter II

The Original Owners and the First Voyage

An act of the British parliament, passed several years before the *Struan*'s time, required that ownership of any vessel constructed or held by citizens of Great Britain or its dominions be divided into 64 equal shares. An owner could hold one share (1/64th) or any multiple of shares up to complete ownership of the vessel (64 shares). No provision was made for a 1/64th share to be further divided. Any partial owner, then, had to hold at least a full 1/64th interest, although such small holdings were not common. Usually 5/64ths or 8/64ths was considered a minimum holding for any owner. A 1/64th interest, for instance, in a typical 1,500 ton new vessel costing $64,000 would amount to only $1,000.

To permit a wider distribution of the much higher costs involved in constructing and equipping larger modern-day vessels, British shipping regulations have been amended since the *Struan*'s time to allow the further division of each 1/64th share of a vessel, so that as many as five individuals or companies may collectively hold a single 1/64th share. Theoretically, then, as many as 320 individuals or companies could be involved in the ownership of a single ship, though 33 of the 64 shares must be held by "persons qualified to own a British ship" as defined by law.[1]

Some six weeks after its actual launching, the *Struan* was officially registered at Saint John on May 29, 1877 as the fifteenth vessel built there in 1877 and was assigned official number 72293. That number, which would remain with the vessel throughout its life, had to be deeply carved into the main beam (the main deck beam closest to the

---

[1] Malcolm Maclachlan, *The Shipmaster's Business Companion*, (Lanarkshire, Scotland: North Sea Books, 1997) p. DO2c.6, DO2d.

main mast) for purposes of positive identification. Even if the vessel were to be renamed, re-rigged, or sold to a subsequent owner, the official number and the signal letters—in the *Struan*'s case S V K Q—would be permanently associated with that particular hull. (In some instances signal letters were changed when a vessel was sold to a citizen of a foreign country).

The initial ownership roster for the *Struan* included four persons, all of Saint John, New Brunswick. George Stymest Robertson, a merchant who had ordered the building of the ship, held 40 shares, or 40/64ths of the new vessel, and was designated managing owner.

John Fraser, the shipbuilder who had constructed the *Struan*, held eight shares or 8/64ths. This arrangement was quite common at the time. The price of a new vessel to the original purchaser presumably included all the costs associated with materials, labor and similar expenditures, but the builder probably expected to realize most of his profit on the project from a continuing stream of income from his share of the vessel's freights during its lifetime.

Similarly, Captain James Cooper, who had been selected to command the *Struan*, held eight shares. Captains' ownership of an 8/64ths share in their vessels was an established way of life at the time. Whether they bought their interests at a discounted price or were awarded the shares as an incentive to operate the vessels economically, efficiently, and profitably for the benefit of the other major owners has not been definitely established.

James Cooper had been born at Essex, SE England, in 1830, and received his certificate of competency as a ship's captain at Plymouth, Devon, England in 1861. He had been a licensed shipmaster for 16 years, and in command of various vessels for 12 of those years, when he became captain of the *Struan* in 1877. He had served as a mate on the *George Marshall* in 1861-62 and aboard the *Great Victoria* in 1863-64. His first command, the *Montmorency*, came in 1864, followed by approximately two years each on the *Charlotte Ann* and the *Empress.* In 1870 he took almost 250 passengers from London to Brisbane, Australia on the *Young Australia* in only one hundred days, but future voyages on that vessel were not so successful. In 1872 he lost the *Young Australia* in a maneuvering accident a few hours after leaving the Brisbane anchorage for the trip back to London.

Lubbock, author of *The Colonial Clippers,* provides some details. "After 10 years successful trading and carrying passengers between England and Brisbane, the *Young Australia* was wrecked May 31, 1872 at the north point of Moreton Island when homeward bound, just four-and-a-half hours after leaving the anchorage of the pilot station. When the ship was in the process of coming about, the wind suddenly died and the strong easterly swell and southerly current set the vessel toward the rocks. The anchor was let go too late, and before it could catch on the bottom, the ship broached broadside to the rocky shore. With some difficulty the passengers were safely landed and the crew soon followed as the heavy swell ground the ship on the rocks. Within a week the wreck had broken in half and was full of water, and a day or two later was sold[2] at auction in Brisbane, after spirited bidding, for a reported £7100[3] to a buyer named Martin.

After the loss of the *Young Australia*, Cooper became master of the *Royal Dane*, an American-built passenger-cargo vessel which had taken 497 immigrants to Rockhampton from London on a 112-day voyage under Captain D.R. Bolt in July-November 1870. After some four years in that vessel, Cooper left to take command of the newly built *Struan* (a cargo carrier only) in Saint John, New Brunswick in the early Spring of 1877. On its very next trip, under the new captain, the *Royal Dane* was wrecked on the coast of Chile while homeward bound with guano[4] in 1877.

The other owner of an 8/64ths share in the *Struan* was John Lane, described on the original registry form as a "master mariner" (as was James Cooper). Lane's interest in the *Struan* can best be explained as an investment on his part, since he was assigned as captain of John Fraser's next ship, the *Munster*, launched in August 1877, some four months after the *Struan*. His continued association with the shipbuilder John Fraser is indicated by the fact that he commanded for about four years a later Fraser-built vessel, the bark *Honauwar*, launched in 1883 for the builder's own account.

When Captain John Lane passed away December 16, 1899 in Hempstead, Nassau County, New York, he had apparently been out of active seafaring duty for about 10 years. In addition to his holdings in four or five New Brunswick stock companies, each formed to

---

2 Basil Lubbock, *The Colonial Clippers*, (James Brown and Son, Publishers, Glasgow) 1921 pp. 120-21.

3 This figure must have included the ship's cargo in addition to the hull itself, since the 10 year old vessel alone would have been worth less than the price paid, even if afloat and in good condition.

4 From the Peruvian "huanu" (dung), the accumulated long-term deposits of seabird droppings scraped from rocky headlands and rookeries on isolated islands, particularly along the Chilean and Peruvian west coast of South America.
Bulk fertilizer in a passenger vessel was apparently considered appropriate cargo at that time. Even a disagreeable or low-rate lading was preferred to coming home in ballast with no freight charges accruing to the owners.

operate a particular steam-powered vessel, he still held an 8/64ths share in the full-rigger *Honolulu*, built by Fraser in 1882. In an ancillary request for probate to the New Brunswick court in 1900, filed to make possible the transfer of property owned there to his wife Catherine (Kate) (Donovan) Lane, the 1/8th share in the *Honolulu* was valued at $1,200, indicating an approximate value for the 18-year old vessel of almost $10,000.

When Captain James Cooper assembled a crew for his first voyage to Liverpool upon taking command of the newly launched *Struan*, he opted for a split set of articles as was permitted at that time by British maritime law.

By putting the mate, the boatswain (bos'n), the carpenter, the cook and the steward on the "by the month" section of the articles, he agreed to pay each of them a specified monthly wage in British pounds for the duration of the trip. These monthly rates ranged from 8 pounds 10 shillings for the mate downward to 5 pounds each for the cook and steward, with the bos'n and carpenter each entitled to 6 pounds monthly.

The 18 seamen (sailors) signed the second section of the articles on a "by the run" basis, meaning that each would be paid a stipulated sum for his services on the trip without regard to how much time the vessel needed to complete the crossing. Whether the ship was at sea two-and-a-half weeks (considered a fast crossing eastbound in the late spring or summer) or more than four weeks (the actual elapsed time), each crew member on the "by the run" section of the articles received at the conclusion of the voyage the set sum he had agreed to before departure from Saint John.

For reason or reasons unknown to this writer, "by the run" sailors (ABs) agreed on signing the articles to be paid 30 dollars, with ordinary seamen receiving only 15 dollars for the trip, whereas the "by the month" hands were to be paid in British pounds.

One might question why seamen would be willing to sign on for a fixed sum without knowing how long the voyage would last. One reason may have been that the captain agreed to an advance against earnings for each member of the crew in the amount of one-half his pay for the trip, such advance to be retained by the sailor or delivered to his family or dependents prior to departure of the vessel. Such generous prepayments of wages were not generally available on voyages where crew members could desert their vessels before their advance wages would be earned. On this initial trip, it was expected that the ship would not touch land until its arrival in Liverpool, so there was really no risk of such a possibility.

The mate and the bos'n must have been known to Captain Cooper, and perhaps had served with him on another vessel, since each was trusted to receive an entire month's pay in advance on his wages before departure from Saint John.

Another factor that may have influenced the sailors to sign on the *Struan* for a "by the run" trip was the fact that the vessel was new and probably would not need extensive painting, major rigging changes or excessive work at the pumps en route—all considerations that insure an easy trip. Although Saint John was a busy seaport, a large number of vessels departing that location were bound, on relatively short-term voyages, for one of the major ports in Britain or in France or Germany; if a sailor wanted to sign on for a two- or three-year trip to Australia or South America, he would be willing to "work cheap" to get to Liverpool or a major European port where many vessels were loading cargoes for almost any destination a seaman might prefer.

It is also relevant that seamen who signed that part of the articles were very well compensated compared to those in the "by the month" category. The Saint John-born bos'n, Michael Cocklan, and the Norwegian carpenter, Christian Hansen, at 6 pounds per month, ended up being paid almost the same amount as able seamen for their "by the run" signing, despite the fact that they would have had substantial responsibilities on a "shake down" cruise by a new vessel.

The mate on this trip is believed to have been a Saint John-born 30-year old mariner named Samuel Thomas Young, though his signature in both places on the ship's papers (sign on and sign off) is illegible. He wrote his certificate of competency number—459—on the *Struan*'s articles when he signed aboard, indicating that he was already licensed either as a ship's officer or master.

Available records do not explain why Captain Cooper engaged such a large crew—24 all told—including 16 able seamen and two ordinary seamen, comprising men from Jamaica, Norway, Scotland, Ireland, Finland, Belgium and the United States. Other crew lists for later voyages of the *Struan*, under both British and Norwegian registry, indicate a manning table usually including, besides the officers, a total of 15 crewmen, but often combining the bos'n's job with that of the carpenter, and sometimes expecting one person to be both cook and steward, though this practice was comparatively rare, since few captains were willing to sacrifice either good meals served on time or cleanliness in their quarters.

There are several possible reasons for the larger than ordinary crew. Captain Cooper undoubtedly wanted to test thoroughly all the equipment aboard his ship under a variety of conditions in various kinds of weather, and may have felt that a few extra hands were desirable in case major adjustments to the sails or rigging, etc. had to be made enroute to Liverpool.

Two of the Saint John newspapers at the time—the *Daily Telegraph* and the *Daily News*—published reports identifying the outbound cargo carried by the *Struan*, but there was little agreement between the two versions of what quantities of which items the vessel was actually transporting. The only commodity listed identically by both papers was 304 tons of pig iron, which, because of its weight, would have been loaded first. The balance of the cargo consisted of deals, deal ends, boards and moldings, put aboard by the firm of Vroom and Arnold, and some twenty tree sections described as piling in one account and spars by the other. With the pig iron loaded deep in the hold to give the ship stability, the *Struan* was able to take on a maximum load, including 219 tons of deck cargo, for which Captain Cooper was assessed at Liverpool the normal extra fee for cargo carried in that manner.

Without the official logbook, which, according to proper authorities, has not been preserved, it is almost impossible to determine details of the *Struan*'s first Atlantic crossing. Other vessels of similar size and rigging style were making the trip in seventeen days, and a few in just fifteen.

It is reasonable to assume that Cooper used some of the time enroute to study the sailing characteristics of his new command under various wind conditions. One of his questions would have been "how close to the wind could his new vessel sail when tacking into a stiff breeze?" Some captains bragged of being able to sail as close as 8° to a headwind, but all vessels, even those launched by the same builder as duplicates of a previously successful model, had different characteristics of performance under essentially similar conditions.

One of the characteristics of a sailing ship was how she answered the helm in a "coming about" or "wear ship" maneuver. "Coming about" (changing from one tack to another in the simplest way while keeping the bow into the wind) was a relatively easy operation if the sailing master was adept at coordinating the work of the seamen handling the sails and yards, the helmsman's steering and his own skill in timing the change of direction through as many as 60°.

A "wear ship" maneuver, on the other hand, involved changing tack through an almost complete circle of as much as 330°, by presenting the stern to the wind. Because this maneuver required more time to complete than did a simple "coming about" operation, and was more difficult to control, it was usually not attempted in restricted spaces or near crowded anchorages.

If ever there was a port that exemplified the old seafarer's maxim, "We sail with the tide," Saint John and the other Bay of Fundy ports are each a perfect example.

In many parts of the world, the difference in water depth at dockside between high tide and low tide is only four or five feet. In the Fundy ports, however, the variation is often in excess of 20 to 24 feet, requiring careful attention by mariners to arrivals, departures, loading, mooring and maneuvering.

Indeed, the incoming tide at Saint John is so extreme that it actually reverses the flow of the Saint John river at its mouth, and causes water to flow upstream over the falls to produce what is known worldwide as the famous "Reversing Falls."

Since well before the late 18th century, small sailing vessels engaged in the coastwise trade, and wood boats bringing fuel and lumber from interior locations would maneuver themselves into Saint John's Market Slip at high or near high tide, and wait for the receding water to rest their hulls on the sandy bottom, where horse-drawn wagons could go alongside to load and unload cargo until the next incoming tide refloated the hulls. If cargo handling had not been completed, the vessel could simply stay in place and go through another tidal cycle.

In a book of autobiographical sketches covering his experiences as Paymaster and Superintendent of Stores for the Military District at Saint John, New Brunswick from about 1879 to 1887, Lt. Col. William T. Baird recorded his observations from the military reservation on maritime subjects. He recalls:

> From the upper windows of our quarters in the stores building, we had a clear view of the [Saint John] harbor from Navy Island to Red Head Battery; and of Partridge Island, Fort Dufferin, the Beacon light and the buoys marking the course of the Bay channel.
>
> During a strong blow from the south, we could watch the sea break over the huge boulders forming the breakwater. Very close to us, gigantic swells upon occasion rose as high as 30 or 40 feet to attack the Ballast Wharf and the railway track. During one terrific gale, a vessel was swept over the rocks and high onto the shore just below the hospital. Another wreck took place in Courtenay Bay itself, though no lives were lost in either instance.
>
> Almost daily we witnessed the arrival of large ships under full

> sail, and other vessels being towed by puffing, plunging gallant little tugs. I have seen in the harbor at one time as many as eight or nine sturdy iron steamers loading with timber, in some cases so deck-laden that I feared for their safety in a storm.[5]

As the *Struan* was towed out into the Bay of Fundy from her berth at the harbor quay on May 29, 1877, Partridge Island came into sight on the starboard side, about a mile offshore from the mainland. The island had been named by Samuel de Champlain for the many flocks of wild partridges he found there during his explorations of the area in the early 1600s.

When the city of Saint John was incorporated in 1785, Partridge Island was selected to answer a provision in the charter for the establishment of a quarantine station to provide some protection for the existing inhabitants of Saint John from the ravages of disease brought by the shiploads of arriving immigrants—in many cases on filthy, crowded, poorly ventilated vessels, usually without adequate food preparation or distribution facilities or any medical care for persons who became sick enroute.

Some ship captains, passage brokers, and even customs agents conspired to evade quarantine regulations, and for a time the situation was so bad that a few ships attempted to discharge their human cargo surreptitiously and escape back to Europe before the captains could be charged and prosecuted for evasion of the Passenger Act.

Beginning about the time of the potato famine in Ireland in 1845, the rush of vessels bringing immigrants completely overwhelmed the facilities on Partridge Island. Where smallpox and cholera had been the main worry of health officials when the station was established, the harried doctors soon found typhus fever, dysentery and measles also rampant on the island.

The city of Saint John itself was not safe, and illness spread as far north as Fredericton. It has been estimated that more than two thousand persons died on Partridge Island or in the Saint John area after being released as cured or having evaded going through the immigration station. Construction work to add facilities at the station came almost to a complete halt, since laborers and workmen would not endanger themselves and their families by working there, even when

---

[5] Lt. Col. William T. Baird, *Seventy Years of New Brunswick Life* (Saint John, N.B.: Press of Geo. E. Day, 1890) pp. 349-50.

substantial bonus pay was offered. Some Saint John natives at the time referred irreverently to "the greenest grass in the world"—meaning the grass growing on the old burial fields of Partridge Island, where the quicklime used in the interment of the epidemic victims provided a constant supply of fertilizer for the grass.

Another landmark that the crew of the *Struan* would have seen as they passed Partridge Island was the steam-operated fog horn, invented by Robert Foulis and placed in service in 1859.

Six or eight miles out into the bay, Captain Cooper, his attention on the strength and direction of the wind, signaled the tug to cast off the tow line, and the *Struan* spread her shiny new canvas for the start of her first voyage.

One of the items of equipment supplied every new vessel was a set of flags, including those of the various countries the vessel might be expected to visit, so that proper "courtesy flags" could be displayed when within the territorial waters of a foreign nation. A set of signal flags was included for identification at sea and for transmitting messages to shore stations, as, for instance, "Require Medical Assistance," "I Need a Pilot," etc. As previously mentioned, each vessel was issued a four-letter identification code upon first registry by the government involved, and that code usually stayed with the vessel throughout its life. The *Struan*'s call letters were S V K Q. The person selected to handle flags at the beginning of a voyage needed to be familiar with the location of the flag locker and his own vessel's call letters, so that he could hoist the correct flags on the signal halyard when required. Although telegraphic messages could be sent to managing owners from major world ports, reporting sailing dates and other important information, there was at that time no telegraphic communication between vessels by radio, or between vessels and land stations. The only record of a ship's movement after leaving port on a long voyage was "ships spoken" sightings recorded by passing vessels. This information would be turned over to the Lloyd's representative at their next port of call to be telegraphed to London, decoded, and published daily in Lloyd's periodicals, providing information to interested persons all over the world. A typical entry might read as follows: "Sighted March 14, British bark *Alexandra Bowdoin* sailing southwest, latitude 22° S, longitude 118° W, sig "all's well" or "Sighted April 11, unidentified white Norwegian four-masted schooner, sailing east, latitude 39° 30' N, longitude 31° 20' W, darkness intervened."

Perhaps the greatest potential hazard to be avoided on the voyage from Saint John to Liverpool was the infamous Sable Island, long known as "the graveyard of the Atlantic." Located some two hundred miles southeast of Halifax, Nova Scotia, and 140 miles almost due south of old Fortress Louisburg, in latitude 40° 56' N, longitude 60° W, the island, composed of shifting sand, has been the site of more than two hundred known shipwrecks and perhaps as many more unrecorded losses.

In the one hundred years preceding the *Struan*'s first voyage, Sable island, constantly swept by strong and irregular ocean currents and subject to dense fog and frequent storms, had been reduced in size by naturally occurring forces from about forty miles in length to less than twenty; from two-and-a-half miles in width to only one; and from almost two hundred feet in height to less than one hundred. Its color and constantly decreasing elevation made it particularly difficult to distinguish from the surrounding seas, except in ideal weather. Ever-shifting sandbars, extending 15 or 18 miles from each end of the island served as traps for the unlucky vessel whose captain thought he had safely passed the hazardous area.

Since 1873, the Canadian government has maintained a system of lighthouses and lifesaving stations on Sable Island. The fact that the lighthouse on the west end of the island has had to be moved on three or four separate occasions to evade the erosion of the shifting sands lends credence to the reports of rapidly changing topography.

Lifesaving stations for a time were fully manned, but over the years have been gradually reduced to a series of safe houses equipped to provide fuel, food, clothing, water and medicine, to which castaways were directed by signs in a variety of languages at intervals along the shoreline. Wireless contact with the mainland since about 1904 has made the reporting of vessels ashore easier and more timely than was previously possible.

Estimated at that time as being over one hundred miles long, Sable Island was known to the early navigators as Santa Cruz. A Portuguese vessel in the early 16th century, whether by intent or by accident, left horses on its shores, and the wild pony descendants of those animals roam freely on the island today.

Captain Cooper was only too cognizant of his loss of the *Young Australia* on Moreton Island, near Brisbane, Australia, in 1872, and it may be safely presumed that he kept carefully away from the well-known dangers of Sable Island.

On his arrival in Liverpool on July 1 (after a passage of a few days over one month), Cooper was shocked by the cries of newsboys hawking papers describing the devastation of Saint John in the great fire of June 20, 1877. Everywhere he went along the waterfront, posters were displayed soliciting funds to be sent to the refugees from the fire and for rebuilding the city. On his return to his vessel after visiting the Merchant Marine office, he carried extra copies of newspapers for his officers and members of the crew, many of whom had homes and families in the stricken city.

The fire, they learned, driven by a brisk northwest wind, had laid waste to almost 40 percent of the entire city. Some 1,600 homes were destroyed. Even stone and brick structures that had previously survived several other serious fires were destroyed.[6]

Believed to have started at George Fairweather's building at York Point, the flames soon became a conflagration, racing before the wind to gobble up frame structures in its path. Pandemonium reigned in the streets. Cartmen usually satisfied with $5 a load for moving furniture and luggage raised prices intermittently as the fire raged and panic-stricken citizens tried to save what they could from buildings in the path of the flames. At the height of the fire, even offers of $35 and $50 a load were being refused. All the schooners and wood boats grounded in Market Slip were destroyed. When the conflagration threatened the jail, a deputy sheriff released the prisoners to take them to a place of safety. Some escaped, but later returned.

Postal employees commandeered a fishing boat, loaded all the mail aboard and sent the boat out into the harbor. Coal valued at $17,000 stored at the gas house caught fire and burned for ten days. Some lives were lost when people loaded furniture and other belongings into small boats and attempted to maneuver them out into the harbor. Many such overloaded craft capsized and several drownings occurred.

Employees of one sailmaking loft on the waterfront dumped all their wooden sailmaking benches into the harbor to save them from the flames; crews in small boats rounded them up two days later. Some of those benches are said to be still in use today.

Knowing that there was no need for a crew to be aboard his vessel at dockside during unloading and loading operations, Captain Cooper arranged to expedite discharge of the entire crew a day or two after arrival, so that those who wanted to return to Saint John could seek berths on vessels bound there as early as possible.

---

[6] George Stewart, Jr., *The Story of the Great Fire in Saint John, N.B.*, (Belford Brothers, Publishers, Toronto) 1877.

Chapter III

Whence the Name?

Many of the wooden sailing vessels built in the Canadian maritime areas before 1900 were commissioned either by individual businessmen/shipowners or by loosely-knit family shipowning organizations. Many merchants, in fact, greatly increased their sales volume and eventual profits by providing equipment, supplies and foodstuffs for the operation of vessels in which they had complete or major ownership. These new hulls were, in most cases, named after the original owner, a member of his family, or a historical figure he particularly admired.

Over a period of some fifty years, for instance, the Troop and Son fleet of Saint John, New Brunswick operated vessels named for just about every Troop in the family. The original head of the operation, Jacob V. Troop, not only had a vessel named for himself but also one for Nellie Troop, Annie Troop, Edith Troop, Kate F. Troop, Josie Troop, Lizzie C. Troop and Mary A. Troop. Their fleet included vessels in which the Troops were majority or exclusive owners as well as ships, barks and other carriers owned by many others who put their vessels under Troop management in the lucrative "China Trade."

Alpheus Marshall of tiny Bear River, Nova Scotia built and named several larger vessels for himself, his wife, and their family members, and put those hulls under Troop fleet management and operation while retaining smaller vessels for operation by his own captains.

Well known examples of sailing vessels carrying names of famous people and places were the *Marco Polo*, the *Florence Nightingale* and the *Cleopatra*.

Some observers claimed they could predict an owner's plans for a newly-built hull from the name the vessel was given at launching. Naming after the major owner or a member of his family, indicated that the ship would probably be owned and operated by the original owners for some time. While it is true that many newly-launched products of British North America yards arrived in Liverpool on their first voyage with ownership transfer papers in the captain's possession, giving him authority to sell, such a predisposition to early sale did not always result in a change of ownership.

In some cases, captains were allowed to decide on their own whether to sell at an attractive price (as was the case when hulls were in demand to carry people and cargo to Australia after the discovery of gold in 1851) or to take advantage of increased freight rates when there was a shortage of vessels.

The increasing use of international telegraph and cable facilities after about 1860 took away much of the captain's decision-making

Bear River, Nova Scotia shipbuilder/shipowner Alpheus Marshall with his wife Tamar, on the occasion of his 83rd birthday in 1906. The *Tamar E. Marshall*, named for his wife, had rescued the *Struan* crew members from their derelict vessel off the west coast of the United States in December 1890. The *Marshall* itself was lost off the U.S. east coast in 1896 under circumstances almost identical to those of the *Struan*.

— *Photograph courtesy Dr. Clarke Fraser*

responsibility by enabling him to receive instructions from owners in Canada from almost any world port within a few hours.[1]

It is understandable that owners who intended to sell their new vessels in Liverpool or one of the other major British ports, probably after having brought over a cargo of "deals",[2] would be hesitant to name the hull for a wife or daughter, since after the sale they would have no control of its type of cargo or operation. Upon sale of a vessel, all rights of the previous owner, including the right to apply for a name change, were extinguished. No subsequent owner could be forced to change the name of his vessel by a previous owner, who may have preferred not to leave his wife's name and likeness displayed on the figurehead of a blockade runner or fertilizer carrier.

For this reason, many vessels intended for sale on the British market upon arrival in the United Kingdom were given innocuous names.

Many of the shipbuilders and shipowners in New Brunswick and Nova Scotia were of Scottish descent,[3] and it is understandable that numerous ships, barks and smaller vessels were launched with place names familiar to the builder or buyer. *Ben Nevis*, *Caledonia*, *Highland Lassie* and *Scotland* are just a few of the better-known Scottish places or people honored by British North America shipbuilders in giving names to their vessels.

Usually, the person who pays for the product has the privilege of naming it. There were some buyers, however, who were so anxious to take advantage of high freight rates that that they didn't care what name a vessel was given as long as it floated. They were also indifferent because official name changes were accompanied by red tape and considerable delay, in the interest of protecting the owners under the

---

[1] Radio receiving and transmitting devices were not installed even on U.S. Navy vessels until 1899. Commercial steamers began to receive the new equipment a few years later, but it was seldom found aboard sailing vessels, probably because of the difficulties involved in recharging batteries on a vessel not having a power plant aboard.

[2] Boards or planks, usually of fir or pine, with a minimum width of about seven inches and a minimum length of some six feet, normally less than three inches thick; probably intended not for use in structural construction, but suitable for being remanufactured after drying into furniture, window and door frames, moldings, etc. by carpenters, shipfitters, cabinetmakers and the like.

[3] Scottish Highland clansmen fighting in support of "Bonnie Prince Charlie's" claim to the British throne were defeated at Culloden in 1746 by government troops, and suffered major casualties in dead and wounded. In 1773 when government agents began to seize the livestock of such surviving Highlanders, a group of them took passage with their families to Canada on the vessel *Hector* which arrived in Pictou, Nova Scotia in the middle of September with about 175 settlers after a rough and difficult six-week voyage involving sickness and shortages of food and water.

original name and the suppliers who may have furnished goods or services under the previous nomenclature.

By April 1877 when John Fraser launched the *Struan* from his yard on Courtenay Bay at the mouth of Marsh Creek in Saint John, freight rates had been forced down severely by a combination of factors that eventually led, even before the turn of the century, to the end of the golden era of wooden sailing vessel construction in British North America.

The surplus of vessels remaining afloat from the fleets hurriedly put together for the Australian trade twenty years or so earlier, together with the growing preference of shippers for the speed and dependability of iron—and later steel—steamships built in Britain and on the continent of Europe, at first gradually, and then with a sudden rush, made the wooden sailing vessel uneconomical to operate.

Shipyards large and small throughout the region saw the handwriting on the wall, and with no orders being received or expected for wooden hulls, began closing down their operations.

Though some of the better financed and more farsighted explored the possibility of producing iron ships in British North America, the absence of any rolling mill to make ship plate and of facilities for constructing engines and boilers convinced even the parliamentary committees appointed for the task of analyzing the situation that iron or steel ships were not a product soon to be built in the provinces. Even as late as World War I much of the tonnage built in eastern Canada for wartime service was still of wooden construction, put together in many instances by hastily trained workers supervised by retired craftsmen called back for service in the emergency.

Since shipbuilder John Fraser and the three initial Robertson owners, George S., James U., and William A., were all of Scottish ancestry, it would be difficult to determine which of them actually chose the name *Struan*.

Recently discovered records reveal that there had been an earlier and smaller Canadian vessel named *Struan*—a 207-ton brig—[4] built in 1846 at Saint Martins,[5] New Brunswick by Jacob Melvin, exclusively owned by James Robertson (perhaps the same James Underhill Robertson who owned a share of the later *Struan* launched in 1877).

---

4 A two-masted sailing vessel square-rigged on both masts. This type of rigging was most often used on craft shorter than about one hundred feet in length.

5 A small community on a protected cove 25 miles east of Saint John, N.B., famous until about the turn of the century for its many shipyards, and home to James H. Moran, who built more and larger vessels there than any other builder.

Documents relating to that earlier *Struan* are interesting in a number of ways.

The official record of its registration illustrates both the frugality of the New Brunswick citizenry at the time, and the shortage of stationery and supplies distributed to officers of the colonial administration.

In 1846, nine years after the accession of Queen Victoria to the British throne, the registrar at Saint John used a form printed in 1834 during the seven-year reign of King William IV (1830-1837) (by lining out in ink the reference to the fourth year of the reign of King William IV and substituting the ninth year of the reign of Queen Victoria). Collector of customs G. Bowger Smith described the vessel as 85 feet long, 22 feet in breadth. He named the captain as Michael Watson and the surveying officer as Charles Dowry. Regretfully, the copy we obtained from the British archives is not of reproducible quality.

John Fraser's family can trace his ancestors back to the 17th century, when an earlier John fought under Dundee on July 27, 1689 at the Battle of Killiecrankie, less than ten miles from the tiny community of Struan.

Struan itself, at the junction of the river Garry with the Errochty, is about forty miles north of Perth, the capital of Perthshire, and longtime favorite residence of the early kings of Scotland. James I was murdered at Perth in 1437 by Sir Robert Graham and his scheming accomplices, including Sir Robert Stewart who expected to have his 72-year old grandfather, the Earl of Atholl, a son of King Robert II, take the throne.

The assassins fled into the countryside, but were pursued and captured by clansmen under the fourth chief of Clan Donnachaidh, Robert "Riabhach" better known as "Grizzled Robert." It is reported that when the young James II became of age, he offered the forfeited Earldom of Atholl to "Grizzled Robert," but the clan chief asked instead that his lands be converted into a free Barony of Struan, which request was granted.

The small churchyard at Struan, Perthshire, has for centuries been the final resting place for members of clans in the district. When I visited Struan in 1991, probably half the stones in the cemetery bore Robertson names. Many inscriptions echoed the respect held for the Struan kirk by Robertsons and Reids and others in the area. Every message on the monuments, regardless of the size or complexity, reflected the dignity surrounding burial in that old ground.

One inscription, on a stone from a 1990 burial, exemplified the tone of the older carvings:

ON THE LOW ROAD I NOW RETURN
TO THE STRUAN KIRK BY THE GARRY BURN
THE LONG SOUGHT LOVE AT LAST TO MEET
WHERE SORROW ENDS AND PEACE IS SWEET.

Another nearby monument echoed the sentiment:

SACRED TO THE MEMORY OF
THE ROBERTSONS OF STRUAN
BURIED HERE FOR GENERATIONS
AND IN LOVING REMEMBRANCE OF
JOHN ROBERTSON
WHO DIED 24 JANUARY 1879
IN HIS 91ST YEAR

Carefully fenced to keep out the sheep grazing on the surrounding hills, the burial ground was for many years kept neat and tidy by the simple expedient of opening the gates to a few animals for a short period of time at regular intervals. This kept the grass cropped short around the stones and markers, without damage being done to the ground cover by overgrazing.

The present church at Struan was built in 1828 on the site of a previous larger stone building believed to have been constructed in the early 17th century. The minister at Struan kirk in 1955 cited early records indicating that the parish was established in the 13th century.

The operator of the tiny Struan Inn during my visit was unable to give any specific information regarding the age of the present building, though he described his establishment as an "18th century coaching inn." The inn had been somewhat modernized, and the two- to three-acre paddock area previously intended for the safekeeping of guests' carriages, wagons and animals, has been converted to a caravan park.[6]

When a single-track railway was built close by the river edge of the property in the 1850s connecting Edinburgh, Perth and a number of other communities with Inverness, a hand-bricked tunnel was built from the paddock area under the rail line to facilitate driving teams to water. Later, the railway added another track alongside the existing one, and a new bridge was built across the rushing Garry, close alongside the original one.

---

6 A caravan park in the British Isles is equivalent to a parking area for motor homes and campers, usually with central shower and laundry facilities.

The Struan Inn, Perthshire, Scotland, circa 1845 before completion of the Highland Railway.

— *Original painting by Orvis Langland; owned by the author.*

Although trains do not stop at Struan today, there was always a station on the site during the early days of the Highland Railway. The Perth Museum has a photograph of the Struan station taken in 1915, noting that the station buildings date from 1898, built to replace the earlier structure which was destroyed by fire in October 1897.[7]

Even a cursory examination of place names in the Scottish Highlands reveals a large percentage that are not of English origin or vintage. Many can be traced to the Celts, the Picts or the Gaels.

Most historians agree that there were Celts in England and Ireland as early as 300 B.C.[8] One author estimates that as late as 1975 there were still about two million persons speaking Celtic in Scotland, Wales, Ireland and some other locations.[9]

A thousand years before John Fraser launched the *Struan* from his Courtenay Bay yard in Saint John, New Brunswick, the Vikings were harassing England and even Scotland. A Danish raid on the mouth of

---

[7] Anthony J. Lambert, *Highland Railway Album* (Ian Allan, Ltd., 1974, reprinted 1994 by Fraser Stewart Wholesale, Ltd., Waltham Abbey, Essex) p. 48.

[8] Gerhard Herm, *The Celts* (St. Martins Press, New York), p. 138.

[9] *Ibid*, coverleaf.

the Thames in 835 was followed by additional thrusts against London and Canterbury about 850.

A few years later, the Viking forces holding London split their army, one group heading north to attack the Picts and Britons in southern Scotland. Before long, large areas in the southwest came under their domination. The invading leader began distributing captured land to his men, and slowly the Viking strategy changed from quick strikes for plunder to longer-term raids to find and appropriate good land for farming.

The last of the Anglo-saxon kings of England, Harold Godwine, was crowned in January 1066, and almost immediately had to lead his army against the powerful Harald Hardrada, King of Norway, at Stamford Bridge near York.

Although the Norwegian king was killed and his horde of Viking invaders soundly defeated, King Harold of England had no time to enjoy the victory.

William, the Duke of Normandy, invaded England in the fall of 1066 with a large French force, and on October 14th near Hastings met the battle-weary remnants and hastily gathered recruits of the army that had earlier trounced the Viking invaders.

Before dark, England's King Harold lay dead on the battlefield, and an almost illiterate French-speaking duke was on his way to being crowned King of England.[10]

After the 10th century sea battle at Hafrsfjord near Stavanger where Norwegian Harald Fairhair decisively defeated a motley collection of minor kings and their followers from various regions of southwestern Norway, a group of the losers fled to the west coast of Scotland and the Hebrides, where they based a reverse Viking operation. Living in the islands, they raided the west coast of Norway at their leisure and fled back to their island bases. King Harald had to put a stop to their piracy to protect his own reputation as a leader, and is reported to have sailed with his fleet to Orkney, Shetland, the Hebrides and even the west coast of Scotland to pursue and eliminate the raiders.[11]

Alexander Carmichael, while still a student of Celtic at University of Edinburgh, in the late 19th century, undertook the task of visiting small out-of-the-way locations in the Scottish Highlands and the islands of Scotland to record old Celtic hymns and incantations. These

---

10 Josephine Ross, *The Monarchy of Britain* (New York: William Morrow and Company, Inc., 1982) pp. 10-13.

11 Gwyn Jones, *History of the Vikings* (Oxford University Press, 1984), p. 90.

A *Struan* oldster, circa 1870.
— *Original sketch by Janet Crichton..*

were orally recited or sung to him by hundreds of willing and cooperative oldtimers, who also provided him with valuable information about both obsolete and dying rites and customs of the Celtic people.

In his later six-volume work entitled "Carmina Gadelica,"[12] are many references to "Struan" with the explanation that "struan" was the Gaelic name given to the cakes or breads baked with great care every year for the feast of St. Michael[13] on September 29, each loaf incorporating all the different grains grown in the area during the year—usually oats, rye, and bere (barley).

Despite the fact that most of the original Robertson owners of the *Struan* were established in Liverpool as shipowners, shipbrokers and shipowners' agents, it is entirely possible that they specified the *Struan* name along with various particulars of construction when they ordered the ship built by John Fraser. Whether or not they were from the Struan countryside or had ever even visited there, the Robertson name alone will forever bind them to the community of Struan as closely as their original registration of the vessel bound them to the ship *Struan*.

[12] Alexander Carmichael, *Carmina Gadelica* (various publishers). Volumes I and II, edited by the author, were published almost coincidentally with his death in the service of his country. Volumes III and IV, edited by his grandson, Prof. James Carmichael Watson, were released a few years later. Volume V, edited by Prof. Angus Matheson, was published in 1954. Volume VI, edited by William Matheson, was completed in 1969 and finally published in 1971.

[13] St. Michael was recognized in Celtic legend as God of the sea.

Chapter IV

Rice First, Guano Later: Thankfully in that Order

Unbeknownst to Captain Cooper at the time, a major change in ownership of the *Struan* took place while the ship was still at sea approaching Liverpool on its initial voyage. On June 28, 1877, at 4 p.m. local time, George Stymest Robertson, who had originally been the major owner with forty shares of the vessel, transferred on the books of the Registrar of Shipping 24 of his shares to William A. Robertson, identified as a Liverpool merchant; eight shares to James Underhill Robertson; and four shares each to Louis Alexander Robertson and Robert Smith, a school teacher from Kircudbright, Scotland. These transactions effectively disposed of all forty of William A. Robertson's 40/64ths share of the *Struan*, reduced his interest to zero, and called into question his right to remain managing owner. The Liverpool *Journal of Commerce* in its July 2, 1877 notice of the *Struan*'s arrival and docking at Birkenhead East Float, gave the owner's name as J.U. Robertson, Saint John, N.B., though on the books of the Registrar of Shipping at that time, he was the holder of only eight shares in the vessel, a 1/8th interest.

From the time the *Struan* reached Liverpool on July 2, 1877 on its inaugural voyage, its harbor movements were carefully monitored by entries in the daily *Journal of Commerce*. First reported as having been berthed at the Birkenhead East Float on its arrival, the ship's dock assignments were followed in each issue of the newspaper.

On July 12 the ship was moved to Canada Dock E; on the 14th to Huskisson North Branch; and on August 7 to Herculaneum Basin. These changes in dockage can be more easily understood when it is recalled that the ship carried two different categories of lumber, some pig iron, and various ship's spars or masts. It is believed that the two lumber items were unloaded at Birkenhead East Float and at Canada

Dock E; the pig iron at Huskisson North Branch; and the masts or spars at Herculaneum Basin, which was a complex of some size, having as one of its major components a graving dock[1] (the 1877 equivalent of a modern-day drydocking facility) where they would be conveniently at hand for installation on vessels being repaired there.

On August 11 the *Struan* was again towed to Birkenhead East Float, where loading for Bombay continued almost until departure on September 15.

When the lumber and deals from New Brunswick had been unloaded at Liverpool and the new cargo for the outbound voyage taken aboard, Captain Cooper set about gathering a new crew. The first to sign on for the upcoming trip was Samuel Thomas Young, the same Saint John native with the scrawled signature who had been mate on the *Struan*'s inaugural trip. The only other member of the original crew to sign aboard was the steward, Leonard Nicholes from Jamaica.

Samuel Lemon, a new young officer from Bangor, Northern Ireland, was hired as second mate. He gave the name of the last ship on which he had served as the *Lillian* of Saint John, N.B. His wage rate was agreed to at six pounds per month, ten shillings less than that paid to the new Swedish ship's carpenter, Allyn Alliser, who had recently been paid off at the end of a voyage on the *Kildare*, registered at North or South Shields near the mouth of the English river Tyne.

From existing records, two things about Samuel Lemon stand out. When he first came aboard the *Struan*, he signed an allotment form authorizing the ship's owners to pay more than 3/4ths of his earnings each month to his dependents ashore. Few crew members in those times took advantage of that opportunity, though the system had been in place for a number of years and was carefully supervised by the British government. Only three other individuals in the 24-man crew participated in the allotment plan on this voyage. Most were in the semi-skilled category, including the cook, steward and carpenter. No ordinary seamen or able seamen chose to use the allotment plan. Unlike other members of the crew, aside from the captain himself,

---

1 A graving dock was constructed by digging a channel into the shoreline usually at right angles to the main waterway, to which a vessel needing repairs could be floated. Fitted gates could then be closed and the water pumped out, leaving the vessel supported on strategically placed blocks and braced in place by timbers installed from the edge of the dock to the outside of the hull at various locations. An excellent photograph of a Norwegian bark in the Hamilton Graving Dock in Belfast, Northern Ireland appears in Michael McCaughan's, *Sailing the Seaways* (Belfast: Friar's Bush Press, 1991) p. 50.

Samuel Lemon stayed with the ship for a number of years, making five or six voyages and progressing up the chain of command to first mate and finally, when Captain James Cooper retired from the sea in late 1885, Lemon was appointed captain by the *Struan*'s new owner.

We have not been able to positively identify the cargo loaded for delivery at Bombay. Based on the fact that Liverpool-Birkenhead[2] were ports well known for the export of coal, however, and knowing that there were very few natural minerals and no deposits of coal in the Bombay area, we can presume that the *Struan* carried coal on the trip originating at Liverpool September 15, 1877.

> "In the Mersey outward bound, ship *Struan*, 1473 tons, Captain Cooper, owner James U. Robertson, destination Bombay."[3]

> "same, Liverpool sailed foreign, September 15."[4]

On November 22, 1877, the ship was spoken at sea and that information eventually telegraphed to Lloyd's in London for dissemination to the maritime community, although the news was not actually published until December 15, ostensibly because the sighting vessel involved took almost three weeks to reach a port where the meeting could be reported.

> "*Struan*, ship, Liverpool for Bombay, 22S and 28W [in the South Atlantic ocean approximately 700 nautical miles east of Rio de Janeiro, Brazil]."[5]

Captain Cooper reached Bombay on January 28, 1878, and spent almost three weeks there seeing to unloading of cargo. History records that there was a serious famine in the Bombay, Madras and Mysore areas between 1876-78, and that as many as five million persons died as a result. Under those circumstances, there would not have been any foodstuffs available for the *Struan* to carry back to Europe.

---

[2] Liverpool, a major seaport in Lancashire, Northwest England, 200 miles northwest of London, is built in a crescent shape along the shores of the Mersey River (estuary) almost directly across from Birkenhead, another port of considerable importance. Since 1886 the two ports have been connected by a tunnel under the Mersey, and since 1934 by a modern vehicular tunnel.

[3] *The Liverpool Journal of Commerce*, Friday, September 14, 1877.

[4] *The Liverpool Journal of Commerce*, Monday, September 17, 1877.

[5] This information, upon being delivered to its subscribers by Lloyd's of London, was copied freely by newspapers at major ports throughout the world.

Accordingly, Cooper sailed for Rangoon on February 28. Three days before docking at that port, 29-year old Swedish able seaman Bent Benson died aboard the *Struan* while the vessel was off East Point. Captain Cooper recorded the cause of death in the ship's records as "dysentery." In accordance with established practice at the time, burial at sea, or committal to the deep, was preceded by a simple service by the ship's master.

Rangoon, Burma's[6] major port, lies on the east arm of the great Irrawaddy River some 21 miles from the sea. The river itself was at that time navigable by smaller craft for almost nine hundred miles further upstream from the city itself.

In all our studies of the *Struan*'s voyages, the one week period (April 6th—13th) in Rangoon devoted to taking on cargo was the shortest loading time by far. The shipload of rice delivered to Hamburg at the end of this voyage could have been loaded in that short time only if enough laborers had been available at the Rangoon docks to keep up an almost constant procession of human rice bag carriers into the vessel, or if slides or loading chutes were in use to permit the sacks to reach the hold with the help of gravity.

The *Struan*, on its homeward voyage, took five-and-a-half months to reach the port of Queenstown, Ireland (now Cobh) whereas the outward voyage to Bombay had required only four-and-a-half months enroute. On August 30, 1878, while about three quarters of the way back to the British Isles, able seaman Phillip Gomez, from Sierra Leone, died aboard ship of what Captain Cooper described in his official log as "Cheab's disease." We have found no reference to any such affliction in any current or turn-of-the-century medical encyclopedia. It may be that Cooper intended to ascribe the sailor's death to "Cheadle's disease," named by a London pediatrician—Walter Butler Cheadle, 1835-1910—to describe the same symptoms he found in children as those of adult patients suffering from scurvy.

The exact cause of Gomez's death is in question, however, because British maritime regulations required the regular distribution of ascorbutics to all seamen aboard ships to prevent scurvy. It is difficult to imagine that a shipmaster would make an entry in his ship's official log designating scurvy as the cause of a crew member's death, particularly when his vessel had been at sea for 11 months and each crew member should have being receiving his daily ration of ascorbutics.

---

[6] Burma has been officially called Myanmar since 1989.

There are even instances on record in which crew members sued a vessel's owners over the captain's failure to provide the required scurvy preventatives during a protracted voyage.[7]

On September 26, 1878, Merchant Marine Office superintendent I. Lyons at Queenstown, Ireland questioned crew members during his investigation into the deaths of Bent Benson (early in the voyage) and Phillip Gomez (near the end of the trip) and certified that "in his opinion, the causes of death of both seamen, as stated in separate entries in the official log are true, and that no blame can be attached to any person on board in either instance." He further noted that the required forms W&E1 (reports of deaths aboard ship) had been filed. As required by law, Captain Cooper deposited with the Merchant Marine Office superintendent at Queenstown the balance of wages due each of the deceased seamen, £5 19s 11d for Benson and £21 18s 1d for Gomez, and turned over Benson's seabag and Gomez's sea chest, all for delivery to their next of kin.

Having received telegraphic orders concerning disposal of his cargo from the ship's owners while the *Struan* was in Queenstown, Captain Cooper sailed for Hamburg and arrived there on October 9 according to endorsements by British consular authorities on the ship's papers (and on October 11 by German customs records). On entering Hamburg harbor, the "Wachtschiff am Jonas"—the office responsible for recording the arrival and departure of ships—made the following notations:[8]

> "Load capacity in cubic meters, 4169[9]
> Number of crew, 25[10]
> Draught in hundredths of a meter, 518[11]
> Agent, Sloman
> Departed November 27, 1878, in ballast, enroute to Cardiff, Wales."

---

[7] Johnson v. the *Karoo*, D.C. Wash. 1892, 49F 651.

[8] This information was furnished by the Hamburg City Archives, Hamburg, Germany.

[9] The *Struan's* original Certificate of Registry rated the ship at 1,473.29 net register tons, indicating that 147,329 cubic feet were available for loading cargo. The capacity in cubic meters was also indicated at 4,169.41 cubic meters, obtained by dividing the cubic feet by the number of cubic feet in a cubic meter (approximately 35.3).

[10] This figure referred to the number of crew members for whom accommodations were available aboard the *Struan*, not to the number actually carried. Twenty-four in the crew, including officers, sailed from Liverpool at the beginning of the voyage. There is no record of any being hired during the cruise, none deserted and two died enroute, indicating a theoretical complement of 22 upon arrival at Hamburg.

When the *Struan* arrived at the Bute docks in Cardiff, Wales on December 18, 1878 to load coal for Montevideo, local newspapers were quoting freight rates on that commodity and destination at 20s 6d (just over U.S. $5) per ton. If the ship had taken aboard its usual coal cargo of about 2,300 or 2,400 tons, then, gross revenue for the first leg of the impending South American voyage could have been as much as $12,000. Welsh coal had a worldwide reputation as an excellent high-quality fuel, and was recognized as entirely suitable for use either in marine or stationary steam boilers. In some markets, its mere identity as "from Wales mines" brought as much as a 6d (12¢) per ton premium over coal from other sources.

By the late 1870s, steam-powered vessels were beginning to take over from sailing ships as more and more vessel owners and cargo shippers realized the advantage of a transportation method that offered a more predictable arrival time. Steam-powered ships, not completely at the mercy of erratic winds, had the additional advantage of being equipped with winches to facilitate the prompt handling of cargo.

To accommodate the increasing demand for coal supplies to fire marine boilers at world ports, some shipping companies and many port authorities found it necessary to establish coaling stations where steam vessels could purchase fuel to continue a protracted voyage.

Many sailing vessel owners then, the *Struan*'s included, found themselves in the position of earning freight fees for taking coal to distant ports, not necessarily to be burned for industrial or household uses, but to be stockpiled for later sale to their direct competitors, the steamships. These sailing ships were, in effect, subsidizing their own eventual extinction, though they continued to carry such cargoes to prolong their utilitarian lives.

Several crew members of the *Struan* hired before leaving Hamburg were discharged on arrival at Cardiff. Some confusion exists regarding the engagement there of 13-year old indentured apprentice William George Hodgson, who, by available records, came aboard on December 21; was carried on the ship's articles as "home on leave" for a week over the Christmas holidays and then discharged December 27.

---

[11] 518/100ths of a meter, or 5.18 meters, is approximately 17 feet, some three feet less than the draft recorded for the fully loaded sugar-laden bark *Guiana* on arrival at Saint Thomas for repairs in 1886. See Chapter XIII, "Ballast Was a Dangerous Cargo." This lighter draft may indicate a less than capacity cargo load. It should also be remembered that the *Struan* was a larger vessel than the *Guiana* by almost 200 net register tons and would therefore not float as deeply in the water as the smaller vessel if loads of equal weight were involved.

It may be that as a 13-year old he did not have the permission of his parents, or Captain Cooper may have overruled the boy's employment by the mate in the skipper's absence. Hodgson was a very determined lad, however, and on a later voyage managed to sign aboard the *Struan*, served out his indenture and was immediately rehired as a fully competent able seaman.

On January 4, when Captain Cooper realized that his first mate held only a mate's, rather than a master's, certificate, he had to appoint a boatswain (bos'n) to fill the manning table in accordance with existing regulations. Seaman Thomas Cook was selected from the crew for that office.

The *Struan* was only about two weeks into its voyage on January 14, 1879 when sailor Edward Anderson was washed overboard and drowned. On March 21, Cooper deposited the ship's articles with the British Consulate at Montevideo, Uruguay and began a stay of more than two months at that port. Before the ship left Montevideo, consular officers conducted an inquiry into the circumstances of seaman Edward Anderson's death. Because the deceased had apparently drawn clothing and/or equipment from the ship's supply chest beyond the amount of wages due him, the records reflect that his personal effects were sold, still leaving a balance due the ship.

On May 22, seaman C.A. Anderson (not to be confused with the drowning victim of the same last name) was transported to the British hospital in Montevideo suffering from pneumonia; a day or two later the *Struan*'s steward, Frank Calema, was also hospitalized there. Both were subsisted at the ship's expense and were carried on the articles as "left sick in Montevideo, unable to proceed."

Ownership of the *Struan* was again affected by the death on May 16, 1879 of William A. Robertson, who had received his 24 shares almost two years previously from George Stymest Robertson. By a will dated September 11, 1876 (before construction of the *Struan*) William left his major assets, which at the time of his death would have included his 24 shares in the *Struan*, to James U. Robertson. Probate was granted on June 24, 1879, and the shares transferred to James on the books of the Registrar of Shipping January 17, 1881.

A complicating factor in the *Struan*'s convoluted ownership involved the fact that James U. Robertson, after the death of William A. Robertson, borrowed £700 at five percent interest from James Nevins and William Welsh, giving a mortgage on all eight of the shares he had received from George S. Robertson on June 28, 1877. Nevins and Welsh recorded a discharge of that mortgage on May 7, 1880, meaning that when James U. Robertson received 24 shares in the *Struan* from the estate of William A. Robertson, he was then the holder of 32 shares (32/64ths) of the *Struan*, making him the major, and probably the controlling owner, since all the other shares were minor holdings scattered among five persons.

Similarly, Louis A. Robertson mortgaged his four shares for a £300 loan from the same Nevins-Welsh duo on October 9, 1879 at a similar five percent interest, but paid off the loan promptly exactly six months later. Again on October 11, 1880, he gave a five percent mortgage on the same shares "to secure an account current" to James Underhill Robertson, shipbroker, Baltic Building, Liverpool. Discharge of this instrument was completed by a receipt for full payment given August 10, 1883, but not recorded by the Shipping Registrar until October 16 of the same year.

Captain Cooper sailed from Montevideo about May 26, 1879 for Callao, Peru, arriving there on August 4 and beginning a four-month stay at that location. We have not been able to determine why the *Struan* stayed in Callao for four full months, until early December. During that time, four crew members deserted and one seaman, Frank Baxter, was left behind at a Callao hospital suffering from disease of the heart and lungs. Whether Captain Cooper was attempting to find a cargo for the return voyage or some other reason kept the *Struan* in port is a matter for speculation.

It is entirely possible that Captain Cooper may have found himself and his ship unknowingly or unwillingly involved in the backwater of the War of the Pacific. In 1879 Chile declared war against Bolivia and soon thereafter against Peru, with whom Bolivia was allied. In August 1879, while the *Struan* was at anchor in Callao, the Peruvian ironclad *Huascar* was captured by the Chilean armor-plated men-of-war *Cochrane* and *Blanco Encalada*. The Chileans, under commander Baquedano, controlled the entire southern portion of Peru, and occupied both Lima and Callao; they did not hand them back until a peace treaty was signed in 1883. While Cooper may have thought neither side would attack a British flagged merchant vessel, he may have "stayed put" until he could better assess the unfolding military situation.

On December 5, 1879, apparently in desperation for a homeward cargo, Cooper began a five day trip to Lobos de Afuera, a rocky island a little over three hundred nautical miles NNW of Callao, where it is presumed he spent the next three months loading guano for the return voyage.

The guano islands, of which the Chinchas were the most famous, were huge jagged volcanic monoliths projecting well above sea level and covered in some cases to a depth of fifty or one hundred feet or more by the accumulated droppings of millions of sea birds over thousands of years. Loading a guano cargo, besides being unpleasant and time consuming, resulted in the ships, the men, and everything in the vicinity being covered by the yellow dust blown about from handling the powdery substance. The best maintained vessel became completely covered with the dust from masthead to waterline during loading (or later unloading) but could be hosed off after the hatches were battened down to retain appearances until the vessel had passed through a rain storm or two for complete cleaning.

An advertisement that had been published in the *Journal of Commerce* July 4, 1877 (while the *Struan* was docked at Liverpool upon completion of its initial voyage from Saint John, N.B.) directed attention of readers to a quantity of guano being offered for sale, and may have gained Captain Cooper's attention and interest in that commodity.

> On account of whom it may concern, This day, the 4th instant, at twelve o'clock, at the Broker's Office. D12 Exchange Buildings West, a quantity of GUANO, about 500 tons (or sufficient to cover rent and charges) of Alta Vela, ex sundry ships, containing about 31.84 per cent of Phosphoric Acid, equal to about 65 to 70 per cent of phosphates. Now lying in a shed situate in Havelock-street Seacombe.
>
> — John Conning & Co., Brokers.

So far as can be determined, the nineteen-and-a-half month Cardiff—Montevideo—Callao—Lobos de Afuera voyage which terminated at Dunkirk July 15, 1880 was the only instance in which the *Struan* was involved in transporting guano.

Chapter V

The Evolution of Maritime Safety Regulations

As early as 1850 the British Board of Trade and the Admiralty began to receive complaints from their own agents, insurance companies, Lloyds' representatives, and some sea captains relating in most cases to overloading vessels beyond their safe capacity. In some instances the complaints involved undermanning (too small a crew to work the vessel safely), or lack of adequate dunnage and bracing to secure the cargoes or ballast in the holds during rough weather.

Others called attention to the use by some navigators of out-of-date or obsolete charts, the increasing frequency of collisions between vessels or between vessels and fixed objects, and to what was addressed as the "growing evil" of drunkenness among seafarers.[1]

In that connection, many captains endorsed the articles of agreement their crews would sign with the words "NO GROG ALLOWED" or "NO SPIRITS ALLOWED" before any sign-ons occurred, in order that every member of the ship's company was aware of the rules. It can only be assumed that any contraband discovered under such conditions went overboard.

We know from various sources that it was the practice of many skippers to lie at anchor for 24 hours or more after they had scheduled their departure and brought their crews on board. That might well have been a simple method for insuring that seamen were completely sober before setting sail.

Most of the criticisms, however, related to the prevalence of improper or dangerous deckloads—the practice of piling huge amounts

[1] Wright and Fayle, *A History of Lloyds* (McMillan and Co., London, 1928) page number not recorded.

of cargo on the vessel's weather deck to squeeze out every last penny of freight charges that could be realized on a given voyage. Deckloads usually involved lumber and/or timber or logs, since these were the commodities that could be transported in the open without being damaged by rain, snow or sea water.

Heavy deck cargoes made vessels less stable and more apt to roll farther in bad weather because of the added weight topside in proportion to that confined below decks (in the hold). The farther a vessel rolled, the greater the danger of capsizing.

Another problem involving deckloads was the difficulty for crew members in getting around on deck, especially at night or in freezing temperatures, to handle the sails when the men had to move quickly over the obstructions trying at the same time not to trip on chains or cables used to fasten the loads in place.

In an emergency caused by a sudden squall, a storm, or an extremely rough sea, lashings sometimes broke or came loose. Then it became a life-threatening problem for sailors to move around over a partially loosened deckload to take in sail, repair damage to rigging, or attempt to stabilize shifting cargo.

A Canadian sailor in 1880 wrote to his family that "we are loaded with lath deck-high, and could not stand much of a brush with bad weather."[2]

It is not recorded whether his voyage ended in tragedy, but it was statements like his, repeatedly put into the hands of maritime officials after a capsizing or loss in a storm, that gradually forced the government to take action.

Even the ships' captains themselves complained. When there was a surplus of capable ships' officers, a conscientious captain who criticized the owner or owner's agent could quickly find himself ashore without a job and perhaps with a reputation as a complainer.

In some cases, captains who owned a share in their vessels were as deeply involved as non-owners, since every extra ton put aboard resulted in more money in the skipper's pocket when voyage profits were distributed. No wonder many owners found it to their advantage to offer shares in their vessels to captains at reasonable rates.

In early 1888, the 16-page blank "Agreement and Account of Crew" forms (the articles of agreement) provided by the government to masters of vessels were being imprinted with a rubber-stamped

---

[2] K.A. Baird, *Letters of a Cheney Settlement Sailor* (in the Saint John, New Brunswick Public Library).

paragraph indicating "it is agreed that the position of the load line disc which now indicates the freeboard shall not be altered."

Instructions to captains for completing these documents also included a notice that "upon every departure from port the actual draft of the vessel shall be recorded in the official log book, and that a substantial fine would be levied against any master who failed to make the required entry." No longer could a captain or shipowner disclaim responsibility for a vessel's overloading. By forcing every captain to observe his vessel from a broadside view to ascertain his actual draft figures adjacent to the Plimsoll disc mark painted on the hull, no skipper could fail to notice whether his load limit disc was partially exposed to view or completely submerged.

Samuel Plimsoll (1824-98) was born in Bristol and started his business career as a London coal merchant in 1853. Fifteen years later he found himself elected a Member of Parliament from Derby. Considered a passionate reformer and known unofficially as the "sailors' friend" he mounted a vigorous and sustained attack in Parliament and in public against errant shipowners while publishing a tract entitled *Our Seamen* beginning in 1872. In spite of an 1874 Royal Commission report to Parliament that "load lines could not be established by law without injury to British ships,"[3] these activities and Plimsoll's dogged perseverance led to the passing in 1876 of the Merchant Shipping Act, making it possible for the Board of Trade to (1) detain any allegedly unsafe vessel; (2) restrict the carrying capacity of cargo vessels; and (3) require on every ship a mark (now known as the Plimsoll mark) indicating the maximum legal load line.

A parliamentary return published in late 1885 analyzing the results of the latest year and the ten-year period since passage of the 1876 Merchant Shipping Act (Plimsoll's safety legislation) brought the benefits of the Act sharply to the attention of the British public.

In the one-year term between July 1, 1884 and June 30, 1885, 28 ships were reported as being defective on account of hull, equipment, or machinery shortcomings. Upon examination, all these were declared unsafe and refused permission to sail until repairs were made. An identical number of vessels was reported in the same year as being unsafe because of overloading or improper loading. Upon examination, all of these, too, were declared unsafe and detained for adjustment of cargoes.

---

[3] A.G. Ford, *Handling and Stowage of Cargo* (International Textbook Co., Scranton, Pennsylvania) p. 159.

During the ten years since the initial passage of the Act, 591 vessels were reported to have had structural defects. Of these, only nine were found upon survey to be safe, and only six of the total were reported to have been improperly detained. Plimsoll found his main satisfaction, however, in the final paragraph of the survey, wherein it was recorded that of the 369 vessels reported since 1876 as unsafe from overloading, only five were found, upon survey, to be "safe to sail."[4] Assuming an average vessel crew of twenty, more than seven thousand men, he opined, had been prevented from sailing under unreasonably and unnecessarily dangerous conditions.

Plimsoll himself was not entirely satisfied with all the terms of the act he had helped promulgate, feeling that the legislation had been "watered down" to appease lobbyists for the shipowners' organizations, and he continued to press for its improvement even after leaving his seat in Parliament.

In its October 4, 1889 edition, *Shippers' Gazette and Lloyds' List* published a transcript of Plimsoll's latest effort in his continuing campaign to strengthen the Merchant Shipping Act. He began his remarks with comments calling attention to and deploring loss of the *Zephyr* which, he argued, had been seen prior to its departure from Cardiff for Brest "loaded to within 17 inches of the top surface of the deck planking."[5]

"In 1876," he continued, "when Parliament acceded to my entreaty to fix a load line on ships, they stultified the value by accepting an amendment of Mr. Norwood's, leaving the owner of the ship the sole authority for choosing where the load line should be painted, and giving him, in addition to this, the power to alter it whenever he pleased. My proposal was that the load line should be fixed by the builder of the ship or some other competent authority. It is no wonder that the load line as it was altered by Mr. Norwood has utterly failed to eliminate overloading. The only purpose for having a load line was to check the overloading practices of certain ship owners; and under these circumstances to leave them the sole authority as to where it should be put, seemed to me at the time, as I stated in the House of Commons, like allowing a draper to determine how many inches he will sell in a yard."

---

4 *The Grimsby Echo and Cleethorpes Directory*, December 9, 1885, page 3.

5 The *Struan's* "Agreement And Account Of Crew" form prepared September 13, 1877 at Liverpool for its first "foreign" voyage (defined as outside the confines of the United Kingdom and the continent of Europe between the river Elbe and Brest inclusive) specified that the distance from the load line to the surface of the weather deck was 5 feet.

Apparently these comments, widely reported in the British press, had the desired effect. On November 28, 1889 the *Western Daily Mercury* of Plymouth (Devon), England reported the case of the master of the steamer *Ardell* from Hull, who had been fined in Cardiff police court the then-unheard-of sum of one hundred pounds sterling. The magistrate commented in open court that he was determined to stop overloading, and that the accused could pay the fine or spend three months in prison, since he had overloaded his vessel by 150 tons, and that the load line marker was completely submerged.

Should one conclude that the offending captain was too severely treated for endangering his ship and crew, the same newspaper only two months later informed its readers that a workman who had stolen one shilling nine pence (about U.S. $.40) from his employer, went to prison for two years at hard labor.

In spite of his having antagonized or alienated a substantial number of shipowners and other members of the close-knit maritime fraternity with his fiery attacks on those who continued to oppose his efforts to make ships safer for those who sailed them, Plimsoll was honored by the launching in 1873 of a 1,444-ton iron ship bearing his name. It was built by the Walter Hood yard in Aberdeen, Scotland at the order of George Thompson and Company, operators of the famous Aberdeen Clipper Line, sometimes known as the Aberdeen White Star Line. Invited to the launching, he attended and watched his namesake christened by the wife of the vessel's captain.

For more than twenty years, the *Samuel Plimsoll* carried passengers and general cargo out to Australia and a few passengers and as much wool as could be packed aboard, back to England from the ports of Sydney and Melbourne, making speed records for some passages that other vessels, even in later times, found hard to equal.

In an ironic twist of fate, the famous ship in 1904, after enduring being sunk in the Thames river, raised, rebuilt, sailed for three or four more years and finally damaged beyond economic repair on a voyage out to New Zealand, was retired as a coal barge at Fremantle, Western Australia.

There, six years after the one-time coal dealer's death, his grimy likeness on the coal barge's figurehead surveyed the mouth of the Swan river and the sea, checking metaphorically for overloaded vessels in the anchorage as the tide and current swung the once-proud hull at its lonely moorings.

I have not discovered how long the Plimsoll ship survived the Plimsoll man, though his mark has been indelibly etched on thousands of vessels throughout the world during the intervening years, and his legacy to the seafaring fraternity of every nation has saved countless lives since he badgered the House of Commons on behalf of all sailors during his early service with that body.

Many of the British shipowners who had objected to Parliament's passage of the Merchant Shipping Act in 1876 did so in the honest belief that other maritime nations would not follow promptly with similar legislation; that British vessels would be limited to carrying smaller cargoes than foreign-registered ships of equal size, and would therefore earn smaller returns on a specific voyage. Though their fears were glossed over at the time by those supporting the load line legislation, delays by other countries in putting similar rules in place proved the objectors at least partially correct.

In spite of publication in *Lloyds' Weekly Shipping Index* almost ten years before the turn of the century, of "mailed advices from America" relating to progress of the load line rules before the U.S. Congress, no real action materialized. Although Captain Bates, the United States Commissioner of Navigation, was quoted as ready to propose "a strong recommendation to Congress for prompt action regarding the establishment of uniform rules governing the loading of American vessels engaged in foreign trade," his concern that the U.S. would be forced "to put into effect a federally supervised system for determination of the proper location of the load line on each ship, or see American vessels eventually subjected to the English rule in British ports and territorial waters" failed to bring the matter to a head.

It could even be suspected that much of the material fed to the British press as "dispatches from America" on the subject may well have been carefully planned propaganda to lead Parliament to conclude that "Uncle Sam was working on the problem." The United States Congress, for the record, waited more than fifty years after Parliament's 1876 action to put into effect equivalent legislation. U.S. merchant sailors were not protected by government-approved load lines until 1929.[6]

In Plimsoll's time, merchant vessels the world over flew the flag of the country to which their owners owed allegiance. A merchant ship owned by citizens of Norway was registered at a Norwegian port and flew the Norwegian flag. Likewise, an English-owned vessel joined the registry at a port in Great Britain and sailed the world's oceans with the British red ensign flying proudly at its mizzen-gaff.

---

[6] Ford, page 159.

Just after the middle of the twentieth century, however, shipowners all over the globe, in a search for means of reducing their tax liabilities, ship registration costs and operating expenses by avoiding restrictive manning requirements and prescribed safety practices put into effect by the major maritime nations, began registering their vessels in such locations as Panama and Liberia. At first just a few owners opted for the newly discovered "Flag of Convenience" method of ship registration. Other shipowners soon found themselves unable to compete with the lower cost FOC classification for more and more marine carriers. By 1989, there were more than four-and-a-half times as many ocean going dry cargo vessels registered in tiny Panama (1,742) as there were in the United States (381). Before long, more than a dozen other smaller nations began to offer alternate registry services.

In some instances, the governments operating these FOC registries made (or make) little or no real effort to enforce meaningful inspections of or periodic maintenance to ships on their registers, preferring to use those repositories merely as a convenient means for collecting fees for which no real or material services had to be provided.

Before and during the Gulf War, a diplomatic problem came to light when there was a high level discussion about whether or not the U.S. Navy, charged with escorting U.S. owned, but foreign flagged, merchant vessels, could legally put U.S. flags at the mastheads of these transports to make clear the fact that they were under protection of the U.S. fleet.

During the first four months of 1996, the Australian Maritime Safety Authority "detained" (kept in port until needed repairs, replacements, adjustments or corrections were made) 89 vessels in various Australian ports for such violations as unserviceable or unseaworthy lifeboats—some with engines that wouldn't even start, much less run; rotten lifeboat accommodation ladders; defective davit control switches; inoperative emergency fire pumps or extinguishers; frozen-up (with rust or paint, etc.) lifeboat release mechanisms; leaking, abraded, worn or cut fire hoses; and nonwatertight hatch covers.

Some of these vessels were able to effect repairs within a few days, but others were delayed for substantial periods of time, especially when funds for major corrections had to be obtained from owners sometimes difficult to identify and occasionally hard to locate.

Plimsoll tried. His efforts to force the British government to endorse the practice of safety within its considerable merchant fleet resulted in the first legally binding regulations relating to safe ship operation by any major maritime nation. His contributions should never be minimized. Under present conditions, however, a worldwide conference and effective contemporary corrective action by 25 or 30 separate and competing governments will be required to bring about any major improvements. The probability of timely joint action in this direction, unfortunately, appears minimal at best.

Chapter VI

The Long Way Around

Captain Cooper brought the *Struan* into port at Dunkirk on July 15, 1880, terminating the voyage he had begun almost two years previously (on November 27, 1878) at Hamburg.

In early August he signed on a new short-term crew for the tow to Liverpool and delivered to vice consul Edward Taylor at the British consulate in Dunkirk all the ship's papers and the official logbook relating to the just-completed voyage.

Samuel Lemon was promoted to first mate, and the tug took the *Struan* into Liverpool harbor on August 13 after a five-day tow, during which the "by the run" crew was engaged in painting and maintenance work which would have been difficult or impossible to accomplish at sea under sail, where all hands could be called away on a moment's notice upon a change in the wind or the need for a major course adjustment. On discharge certificates for sailors on that short trip under tow, Captain Cooper took the opportunity to assign performance ratings to members of the crew—apparently the first time he exercised this practice.

The ship's cook, steward, bos'n and mate were all given a "very good" on their discharge certificates, but only two of the able seamen—one of them over fifty years old—received a rating better than "good." That age was considered at the time to be near the end of the normal working life of a seaman aboard a sailing vessel. The constant requirement for climbing into the rigging to handle sails in all kinds of weather had a tendency to age sailors prematurely, though there were exceptions. On one voyage, the *Struan* shipped a 64-year old able seaman who apparently was able to perform his duties without difficulty or criticism.

The contemplated destinations given on the cover sheet of the crew list for the *Struan* voyage begun September 23, 1880 and the actual route selected as the vessel progressed on it voyage amazed me almost to the point of disbelief.

Why, I wondered, did the *Struan*, bound for Karachi, India from Liverpool, set off to sail around the Cape of Good Hope at the southern tip of Africa, rather than taking the more direct course through the Straits of Gibraltar, the Mediterranean Sea, the Suez Canal (which had been opened, although not entirely completed, in 1869), the Red Sea and the Gulf of Aden to the Arabian Sea?

After considerable research, several answers presented themselves. For some time, shipmasters utilizing the Suez Canal had been sending back reports of "vexatious quarantine regulations" in effect there. Mariners may have assumed there would be delays before or during transit, or even that payoffs to officials would be expected or required to expedite passage. There was also a generally accepted recognition by members of the Maritime fraternity that the Suez Canal was better adapted to the transit of steamships than of sailing vessels. Steamers could move through the canal under their own power, saving tons of expensive coal as a result of having chosen the shorter route. Sailing vessels, on the other hand, had to be towed the entire length of the waterway, adding substantial voyage charges for tug charter.

Another factor was that the canal had not yet been lighted; there was no illumination to identify shorelines at night, contributing to the fear of a possible grounding or collision, since passage through the 104-mile canal required from 18 to 24 hours, and could therefore not be completed during daylight, especially during winter months.

Fears by some fleet owners of hostile military action in the area were not realized until more than a year later, when the canal was in fact closed before the battle of Tel-el-Kebir. If, for whatever reason, the canal had been out of service for a substantial period of time, a captain who had invested 45 or 60 days getting there it would have been faced with the choice of retracing his inbound course and sailing around the tip of Africa, or riding at anchor for the length of the delay, risking a shortage of drinking water and food in an area where replenishment might have been difficult or even impossible.

The strongest argument for some against using the Suez Canal at that time, was the regulation that ships travelling in one direction were required to tie up to mooring piles (provided at frequent locations along the route) when a vessel travelling in the opposite direction approached.[1] Although that rule may have been instituted to avoid the

---

[1] David McCullough, *The Path Between the Seas* (Simon and Schuster, New York), pp. 345, 346.

danger of collision in the narrow waterway, it was undoubtedly a time consuming and frustrating impediment to efficient use of the canal.

Another concern was that there was no real provision for dealing with power vessels that lost mobility in transit. There were few, if any, places in the entire length of the canal wide enough to accommodate a disabled steamer so as to keep it out of the path of other traffic.

Even steamships of considerable size prior to about 1900 had no facilities for radio communication between ships or even for ship-to-shore use. So, when a breakdown occurred, there was an inevitable delay in notifying canal operators of the problem, and a further substantial loss of time before a tug dispatched from whichever end of the canal was closer could reach the stricken craft. Retrieving a vessel blocking the canal often required that other traffic be held at one or both entrances in order to give the tug ample room to operate without danger to other passage-takers. A single vessel blocking the canal could easily have closed it to traffic for two or three days.

In early 1882, "Maritime Intelligence" columns in newspapers around the world carried telegraphic dispatches from the Suez Canal area describing two such blockages within a month by steamships whose power had failed, steering had malfunctioned, or some other mechanical or navigational mishap had left them dead in the water, out of control, or aground.

> "The British steamer *Harker*, enroute from Simonisaka [Simonoseki, Japan] to the United Kingdom, has stopped the traffic in the Suez Canal."[2]
>
> "The Glasgow registered steamer *Strathmore*, from London for China, has stopped the traffic in the Suez Canal; disabled."[3]

Given the frequency of blockages due to vessel breakdowns in the Suez Canal, it is no wonder that engineers designing the later Panama Canal made the decision to draw vessels through the waterway by land-based railway locomotives under full control of the canal operators. No vessel was to be permitted to use its own engines during transit in critical areas.

Finally, fees charged for passage through the canal were substantial. The French firm which held rights to operate the waterway reported in 1885 to the owners of its 228,000 shares a net profit for the previous year of 35 million francs.

---

[2] The Dublin *Freeman's Journal*, Friday, April 21, 1882.

[3] The Dublin *Freeman's Journal*, Saturday, May 19, 1882.

While waiting outside the Liverpool Merchant Marine Office on September 21, 1880 for his turn to formally engage his new crew , Captain Cooper had time to inspect one of the notices posted there for the information of all British seafarers.

The government of the United Kingdom in 1856, largely in response to complaints by sailors' dependants, implemented a system allowing seamen to deposit their wages where they actually received the money. British law provided that sailors had to sign aboard and sign off their ships in the presence of a shipping master or his assistant in a merchant marine office at ports in Great Britain and its colonies, or before a British consular official in foreign countries. This new rule made it possible for them to surrender their wages to be held on deposit or sent to their families or other designated persons before it was lost or "spent in frivolity." Two-foot by three-foot signs were conspicuously posted at marine offices, advising that:

> The Board of Trade, being desirous of encouraging seamen to save their earnings and prevent as far as possible the risk of their losing their money after being paid off from their ships, have established seamen's savings banks. Wages may be deposited or remitted free of expense from this port to their friends residing in or near any port in the United Kingdom. Free particulars may be obtained from the shipping master.

The *Struan* was "spoken" by another vessel off the southwest coast of Ireland (in latitude 50° N, longitude 15° W) on October 1, just a week into her voyage. Apparently the vessel reporting the sighting was outbound from the British Isles, since her meeting with the *Struan* was not reported and telegraphed to London until October 22.

It was not until November 6 that the *Struan* was again observed at sea. In latitude 4° N, longitude 27° W (in mid-Atlantic about half-way between Brazil in South America and Senegal in Africa) she was spoken by a passing vessel, but the report did not reach London until almost six weeks later, on December 17. By that date, the *Struan* was probably very close to making her passage around the Cape of Good Hope at the southern tip of Africa. Almost another three months were required for the northward voyage through the Indian Ocean, past Madagascar, across the equator northbound, all the way to the north-east corner of the Arabian Sea.

The *Struan* arrived in Karachi, India on March 5, 1881. While in harbor there, Captain Cooper and an able seaman named Emile

Walker agreed by mutual consent to the seaman's discharge at that port. He was paid off and put ashore with his belongings and his £7 accumulated wages.

Cooper arranged for his ship to take on ballast, picked up the ship's papers from the port authority office on April 8, and the next day sailed for Newcastle, New South Wales, Australia, arriving on June 13. On the 14th, shipbroker and commission agent Robert B. Wallace, after conferring with the ship's officers at anchorage, filed an "inward crew list" with port authorities enumerating the 23 persons aboard the *Struan*,[4] and the vessel took her place in line to load coal at Dyke wharf.

When we first inspected the crew list prepared at Newcastle, it appeared that one or two individuals named on it were not listed as crew members on the "Agreement and Account of Crew," commonly known as "The Articles." In our effort rationalize the two lists we almost missed some very interesting information contained in the report.

Captain Cooper's wife and daughter accompanied him as passengers on that voyage! The passenger column of the subject document includes only one entry—"Mrs/Miss Cooper." This suggests that they may have sailed on other trans-oceanic journeys either before or after that voyage, although no other records we have examined included similar information.

It took about a week for the *Struan* to discharge her ballast in the outer harbor designated by the harbormaster's office, before she was hauled in alongside No. 7 hydraulic crane to load Wallsend coal for San Francisco.

In the meantime, Captain Cooper had been busy with a court case involving a dispute concerning wages claimed by one of the *Struan*'s sailors. The facts are carefully laid out in the records of police court activity published in the *Newcastle Herald*, June 25, 1881, page 8.

Several members of the ship's crew deserted while in Newcastle harbor, but by the time the *Struan* left there on July 6 with 2,177 tons of coal, only one had been apprehended. In police court on June 29, seaman Antonio Walker, "for deserting from the ship *Struan*," was "sent to gaol for 14 days."[5]

The *San Francisco Chronicle* on October 16, 1881 announced the ship's presence in the Bay area:

---

4 Australia Archives, CRSC667 Crew and Passenger Lists, Port of Newcastle, Inwards: Roll 15—December 1880 to August 1881.

5 *The Newcastle Herald*, Thursday, June 30, 1881, p. 2.

NEWCASTLE POLICE COURT

Friday, June 24th, 1881

(Before Mr. Buchanan, P.M.)

WAGES CASE.--Martin Jennings, a seaman of the ship Struan, sued Captain James Cooper for £7 2s alleged to be due him for wages. Mr. Dart appeared for plaintiff; Mr. Brown for the defence. The evidence shewed that plaintiff, who bore a good character aboard and had never intimated that he was sick, complained on recently arriving in Newcastle that he was unable to work, and wished to see a doctor. Captain Cooper accordingly took him to Dr. Page, who pronounced him to be suffering from a stricture of long-standing. The captain arranged to have him examined three times before deciding, and if then found unable to proceed to sea promised to have him admitted to the Hospital. Meantime plaintiff absented himself twice without leave, and obtaining a certificate from another medical man, claimed to be discharged, although told that if he consulted a second doctor it would be at his own expense. His Worship held that there was no case: plaintiff had no right to sue, for he had absented himself, as shown by the medical certificate, without there being any "reasonable apprehension of danger to his life," as required under the Act. It seemed that Captain Cooper had done all that a kind, prudent, and thoughtful officer could do in the matter. The case was dismissed, and 21s professional costs ordered to be deducted from the balance of wages due when correctly calculated.

Newcastle Herald—Newcastle, New South Wales, Saturday, 25 June, 1881, page 8.

> Arrived, the British ship *Struan*, Captain Cooper, 101 days from Newcastle, N.S.W., 2177 tons coal for Geo. J. Theobald and Co.

Many Californians in the 1870s and 1880s considered the San Joaquin valley as the "breadbasket for northern Europe." San Francisco newspapers heralded the assembly of a great fleet of sailing vessels in the area to load a single season's grain harvest in 1881.

The *Daily Alta California* on Wednesday, November 23, editorialized:

> It is safe to say that this month will be the greatest for wheat shipments that we have ever had. So far, 50 vessels have cleared and there are still eight days to hear from. Freights are stiff and ships are again holding out for higher [freight] rates, and they will get what they want before long.

After having its coal cargo unloaded and the vessel's interior cleaned, the *Struan* was towed to Port Costa to load wheat on November 20. A little over a week later, with its wheat cargo stowed aboard, the *Struan* cleared for Cork and sailed on December 3, carrying 42,586 centals[6] of wheat (4,258,600 pounds or 1,901 tons). Freight rates to British ports in effect at the time were £3 2s 6d per ton. At the $4.88 per British pound exchange rate, the 1,900 tons at slightly over U.S. $15 per ton, would have provided a $28,500 fee for transport of that cargo alone.

Even without knowing the nature of the cargo or its applicable freight rates on the first voyage to Karachi, it is evident that the total freight earned for the Karachi delivery from Liverpool, the coal cargo from New South Wales to San Francisco at an estimated 24s per ton (about U.S. $6) or some $13,000, and the wheat from San Francisco to a British port, could easily have exceeded the original cost of the vessel itself when built four years earlier.

It is interesting to note that although Captain Cooper listed his destination as Cork, Ireland when he sailed from San Francisco, he stopped at Queenstown (now Cobh) for orders, and ended up unloading his cargo at Dublin, having arrived there on April 22, 1882.

---

[6] A name for a weight of one hundred pounds of a commodity, especially grain, suggested in 1859 by a lawyer named Danson, but not formally legalized for use in trade by the British government until about 1880.

Chapter VII

The Wooden Ship Era

A number of circumstances in 19th century New Brunswick and Nova Scotia combined to provide an almost perfect setting for development of a wooden shipbuilding industry based not just on a few major yards in the larger ports, but on dozens of launching ways in small yards scattered over hundreds of miles of coastline. The ready availability of timber, situated near easily established shipbuilding sites, and a ready and tax-free market in Great Britain for completed vessels and the timber cargoes they could carry to sell in Liverpool or one of the other large British ports, provided a "made-to-order" opportunity for a spectacular surge in tonnage launched.

Any convenient location on either side of the Bay of Fundy, or almost any creek or river near it, was a suitable site for the construction of a vessel. The extreme Fundy tides (20-foot differences between high and low water were common) backed up many of the coastal streams and rivers at high tide to depths that would accommodate launching hulls of substantial size.

Anyone with a minimum amount of capital who owned or controlled the use of such a site had only to obtain the services of an experienced master shipwright to be in a position to build ships. Timber of many different species was available on the thousands of acres of forest adjoining the coast, and a constant labor source was guaranteed by the continuing and seemingly ever-increasing stream of immigrants to these two eastern Canadian destinations.

Many of the established shipbuilding ventures had begun on a limited scale even in the 18th century. Small yards in those days turned out coastal schooners and similar smaller vessels, most of them manned by crews from the tiny communities where they were built. As shipowners and operators continued to call for larger and larger vessels after about 1840, these yards expanded their capabilities until most of them could launch a ship or bark of 1,000 tons or more, usually competitively priced and in most instances constructed to the same standards of quality and workmanship found in the larger yards of the major BNA ports.

British North American built vessels were referred to until about mid-century as "plantation built" in official government records and regulations. Although there were serious early criticisms of some of these hulls, mainly because of the use of unseasoned softwood timber and evidence of less than expert workmanship, this situation changed when the offending builders found that they could not recover their costs in an international market where good quality hulls were easily available.

When Lloyds' representatives and surveyors were assigned to major BNA ports after about 1850, quality of construction improved dramatically, since Lloyds' rating of A4, A6 or A7 (class A for 4, 6, or 7 years) became necessary to guarantee either the salability or the insurability of the vessel.

Even an A10 rating from Lloyds', of course, did not warrant a ship's safe operation for any specific length of time. Many a well-built vessel was lost on its first voyage through no fault of the builders or owners. Some vessels, on the other hand, had particularly long lives. Twenty, 25, or even 30 years were figures not unheard of for vessel longevity.

The generally accepted premise was that a BNA wooden hull built in the 1840s had an average life span of about 10 years. This increased to around 13 years in the 1870s and to almost 15 years for ships constructed just before 1890 at the height of the wooden shipbuilding frenzy prior to the advent of the iron or steel steamships. Such generalities, however, bear examination. For instance, number of years in service by a sailing vessel was usually calculated as the time between its launching and subsequent loss by fire, sinking, stranding, failure to arrive, etc., or disposal to an owner in another country. Once transferred to foreign registry, however, it was difficult to monitor its time in service under a different flag (sometimes even with a different name) and under later third-country ownership. Since many 8- to 12-year old BNA hulls were sold to and operated for years by Norwegian and other flag interests after their initial service under the British flag, the actual average life figures would have been substantially extended if *total* service time was considered.

While these improvements in vessel longevity were perhaps mostly due to closer and more frequent inspections by Lloyds' representatives after 1850, the improvement in construction methods and efficiencies also improved as a result of negative feedback concerning the failures and weaknesses in a larger number of vessels in the world fleet built by BNA yards. The increasing use of iron knees in place of wood, and strap iron "diaphragm bracing" under the hull planking to reduce "working" or movement of the joints as a vessel labored in rough seas, were also factors. In many instances, this iron was 3/4 to 7/8 inches thick by as much as five inches wide, and 30 feet in length. Installation required that the vessel's ribs be strong enough to retain their needed strength even after a one-inch deep "dap" was cut for the bracing to be "let in" under the planking.

Just as in homebuilding today, BNA shipbuilders in the 1800s were building for two different markets. A "speculative" builder might launch a hull designed for quick sale to one of Liverpool's busy ship brokers, while a "contract" builder would often have an experienced shipmaster employed by the purchaser constantly looking over his shoulder.

The poor reputation of BNA vessels in the 1830s and 1840s ended once the offending builders had "cleaned up their act" or gone out of business. The criticisms of British captains who insisted that all BNA hulls were "unsafe to sail in or sleep on" gradually diminished as the quality of the vessels substantially improved, though their opinions contributed to prices of BNA vessels at British ports through the mid-60s (except for one to two years when ships were at a premium for transportation of passengers to new gold discoveries), being no higher than prices in effect as far back as 1825, when deficiencies in quality of construction had been a major problem.

Wage rates paid by builders highlight the interdependence of shipbuilding and farming in the 19th century maritimes. Most records that have become available indicate that almost no shipbuilding was done at the outlying smaller yards in the spring of the year when the family men needed to remain at their homes to plant their crops. The highest wages ($1.25 to $1.50 per day) were paid in the summer when good weather and long hours of daylight permitted maximum productivity. Once their crops were planted, farmers could augment their incomes by going to work in the shipyards.

Fall wages were reduced 15 or 20 cents per day below the summer rate, and winter rates fell to about two-thirds of the applicable daily summer wage, apparently dictated by the easy availability of subsistence farmers who could take full-time employment in the winter months away from their farms. Some, however, earned additional income by cutting lumber from their own property and hauling it to the nearest shipyard to be sold for planking, decking or ceiling, at an agreed rate. Some farmers were encouraged to search during the winter for trees on their own holdings that would provide knees for vessels to be built during the following year.

A book published in 1990 shows that the 1881 Canadian census figures for New Brunswick indicate an average annual wage of $241 earned in 1880 by each census respondent who identified himself as a shipyard worker.[1] At the $1.40 per day basic wage prevalent in outport yards at the time, this would suggest that each man worked about 172 days during the year at the yard and the balance of the time on his farm or at another trade. This figure does not take into account the farmers/shipyard workers who, because they worked less than half the days in a year at a shipyard, may have classified themselves in the census as farmers rather than shipyard workers. Such a choice by any appreciable number of persons could skew the results of such a study to a substantial degree.

Among builders of wooden ships in 19th century Canada, as elsewhere, there were differing opinions on the suitability of materials used. Although the principal classification agencies (Lloyds' and Bureau Veritas) generally dictated the species of timber to be used for various purposes in construction of a vessel if it were to be rated "A" for four or seven years, builders were given almost *carte blanche* in their selection of the metal used for bolts, drifts, and fastenings.

Many shipowners and builders agreed that every hull should be constructed with nonferrous (not iron) bolts and fastenings to well above the normal water line. This was based on the premise that salt water did not affect fastenings made of copper, brass, or Muntz metal, and that these would therefore hold securely for the entire life of the vessel without needing inspection or replacement after a few years. Pure copper had been found to be too soft and thereby unsatisfactory in resisting bending and shear strain after installation, particularly in larger and/or longer vessels.

But not all shipbuilders and ship repairmen agreed. Muntz metal (60% copper and 40% zinc), invented in 1832, was harder and held a thread for a nut better than copper, but it still lacked the rigidity or resistance to bending that iron afforded.

More than one old-timer made no secret of the fact that he had seen two- or three-year old hulls put into drydock for substantial repairs where copper or Muntz metal (brass) fastenings were so loose in their holes that the drift, dowel, or bolt itself could be pulled from

---

[1] Eric W. Sager with George E. Panting, *Maritime Capital—The Shipping Industry in Atlantic Canada, 1820-1914* (McGill-Queen's University Press), 1990, p. 183.

the hole with bare hands. Part of the problem, they insisted, stemmed from the fact that it was almost impossible to drive soft metal (nonferrous) fastenings into net-measurement holes without a few of them bending and "going astray" as they were driven. For that reason, some builders began drilling holes for nonferrous fittings a minute fraction oversize, which, while making installation easier, greatly reduced the fastenings' resistance to movement when the ship was under working strains in heavy seas.

Some other maritime operators were adamant that only iron offered the strength and shear resistance that was necessary in constructing vessels. These early-day shipbuilders argued that a plain iron bolt or drift driven tightly into a net-measurement hole would rust over its entire surface, creating a tight joint far stronger and more resistant to movement than a nonferrous fastening could hope to accomplish.

As facilities for galvanizing[2] became more available in the latter part of the 19th century, many vessels were fastened with galvanized iron and later with galvanized steel drifts and bolts. The thin coating of zinc imparted by the galvanizing process very slightly increased the diameter of the individual fastening, contributing to a very tight fit in net-measurement holes.

Shipbuilders who had purchasers specifying nonferrous fastenings were of course willing to accommodate them, but charged more for the expensive nonferrous materials.

It is interesting to note that some contract builders who normally used one type of metal fastenings in vessels for certain fleet operators would use the other type when building for their own account or for other owners with different specifications.

Until the mid 1840s, when almost all the vessels built in British North American yards were still in the under six hundred net register tons category, masts were usually fashioned from a single tree. Historians have recorded the experience of one shipbuilder whose construction ways were at a remote location, but within sight of his farmhouse.

When it was time to shape the masts and spars for the *Harriet Campbell*, the builder employed an expert sparmaker who brought his adzes, draw knives, roughing planes, and other tools to the site of the shipways where, provided with room and board in the shipbuilder's

---

[2] A process by which iron or steel objects are first "pickled" in an acid bath to remove mill scale and rust, rinsed with water, and then immersed in a vat of molten zinc, giving them a thin rust-resistant coating.

home, he spent 15 weeks tapering and finishing the required masts, yards, and other spars from logs previously stockpiled by the shipbuilder. For that exacting work, sparmaker John Smith was paid at a daily rate equivalent to a 50% premium over that in effect for ordinary shipyard labor.[3]

As vessels increased in size, masts had to be longer, larger, and stronger, and it became impractical or even impossible to find suitable trees of sufficient size and quality for the purpose. Accordingly, shipbuilders adopted the practice of equipping vessels with sectionalized masts made from two or even three separate pieces, girded by heavy iron straps at the lapped joints. By the time the *Marco Polo* was launched in 1851, the lower main mast section on a 1,500 ton vessel sometimes exceeded 36 inches in diameter.

Much has been written about the relative costs of labor and the material components for a timber vessel built in New Brunswick or Nova Scotia at various times and at different locations during the 19th century. While it is easy to understand that the twenty-odd yards in Saint John and its environs bought most of their timber from established timber merchants, dealers, or contract loggers, some of the small shipbuilders at more remote sites supplied most of their needs by sending their own crews to fell and trim appropriate trees at timber camps on their own property.

While such activities perhaps reduced their paid-out costs for construction timber, it added substantially to the labor costs involved, particularly if the logs had to be "whipsawed" into planks by hand. Although it was theoretically possible to build a vessel using no mechanical or labor saving devices, the need to compete price-wise with other builders necessitated whatever labor saving practices were available.

By 1849 a number of steam-powered sawmills in the Saint John area were producing ship planks. The sawyers, who had previously earned their living by laboriously cutting planks by hand with their whipsaws, attacked the power sawing of ship planks as a "pernicious practice" and organized to "remonstrate against this unjust and odious evil and the misapplication and prostitution of steam power for purposes of monopoly."[4]

While there are few records of any serious movement to organize labor unions in the widely scattered smaller yards in outlying locations,

---

[3] Charles A. Armour and Thomas Lackey, *Sailing Ships of the Maritimes* (McGraw-Hill Ryerson, Ltd.) p. 115.

[4] Richard Rice, *A History of Organized Labour in Saint John, New Brunswick, 1813-1890* (MA Thesis, University of New Brunswick, 1968), pp. 24-25.

there were, during the late half of the 19th century, frequent attempts by the various shipbuilding trades in Saint John to form unions for the benefit of their members. In many cases, the ups and downs of the economy, falling freight rates, harsh winters, a surplus of available labor, and the disruptions caused by the great fire of 1877, seemed to combine to hinder these efforts. Publication in 1877-78 by Saint John and Halifax newspapers of reports detailing the failure of ship carpenters in Scotland to achieve improvement of their wage scale undoubtedly had a dampening effect on any surge of enthusiasm for a work stoppage to promote pay increases in the Maritime provinces.

The Scottish carpenters went back to work, after a six-month strike, for exactly the same wages they had been paid prior to their walkout—eight pence per hour, or about U.S. $.16—for a ten-hour day. Since many shipyard workers in New Brunswick and Nova Scotia already earned almost that wage, particularly for summer work, they saw no benefit in being off the job for a long period with little chance of improving their position. As late as 1883, shipwrights along the River Clyde and at Scotland's smaller outports were being paid 36 shillings a week—probably six days—still about $1.50 per day.[5]

When the wooden shipbuilding industry collapsed after the 1880s, there was little need for ship carpenters, shipwrights, ship joiners, riggers, caulkers, shipsmiths or sailmakers, and therefore little call for unions to support them.

The one waterfront group that had been able to maintain (with fluctuating rates of success) any real sense of organization was the Ship Laborers' Society. The year 1849 saw the end of the "truck system," whereby labor involved in loading and unloading vessels was paid for in "chits" or "script" on the stores operated by the merchants who also owned the vessels.[6] When the workmen finally realized that merchandise purchased for a one dollar "chit" at the store probably cost the merchant only 50 or 60 cents, they concluded that their pay was only a little more than half what they had thought. Payment in currency became the basis for all future work.

After years of wrangling with each other over a myriad of minor differences, the Saint John Ship Laborers' Society and the local shipping agents and vessel owners finally agreed in 1887-88 on a workable plan to reduce friction between the parties. On their part, the laborers agreed to:

---

[5] Alexander Stephens and Sons, Ltd., *A Shipbuilding History 1750-1932* (Ed. J. Burrow and Co., London, 1932).

[6] *Ibid.*, p. 33.

1. Permit the merchants, shipowners and stevedores to select specific individuals from the union roster for their work gangs. (The union would no longer insist that each gang include a number of slow or incompetent workers).
2. Allow seamen in regular ship crews to assist in or actually perform loading and/or unloading operations without union objection.
3. Withdraw their objection to the use of steam winches for handling of cargo.[7]

The owners, agents, operators and stevedores undertook to:

1. Employ only members of the union.
2. Contract only with stevedores who employed members of the union exclusively.
3. Use "every legitimate means" to induce captains and other owners to follow their example.[8]

Although not generally recognized as such, the Ship Laborers' withdrawal of their previous objection to the use of steam powered winches for handling cargo may have been a major contributing factor to the rapid decline in construction of wooden sailing vessels. Such ships had historically been loaded and unloaded by using teams of horses working on the docks to handle the hoisting of cargo. Even the early iron steamers, however, were equipped with steam powered winches. Such equipment, though somewhat "jerky" in its operation, speeded up cargo handling to the extent that the number of days required for loading or unloading a vessel could be reduced appreciably, making possible faster turnaround times and more revenue producing voyages per year.[9]

Some sailing vessels in the later 1800s were equipped at launching, or retro-fitted, with donkey engines and boilers in an attempt to equal

---

[7] The union's objection to the use of steam winches was based on the perceived danger to its members in working on vessels so equipped; instances of death or injury to cargo handlers from erratic operation of the steam powered winches were offered to substantiate the "hazard" claim.

Before this agreement, cargo handlers on steamers had received a daily wage higher than that paid for work on vessels not equipped with steam winches.

[8] Rice, p. 107.

[9] The Norwegian full-rigger *Struan*, for instance, arrived at Port Townsend, Washington on October 8, 1890 in ballast from Iquique, Chile, after a 40-day voyage. Loading, or waiting to load a lumber cargo destined for Australia, however, required 44 days.

the cargo handling efficiency of the steamers, but these installations and their fuel supplies required substantial space and thereby reduced the capacity for carrying deck loads.

Even as late as the mid-19th century, any British shipowner choosing between an iron or a wooden-hulled vessel needed to consider the government's regulations regarding transport of the mails. If he hoped to obtain, or even retain, a mail-carrying contract in connection with a passenger-only or passenger-cargo service between the home country and Australia, India or Canada, for instance, he was forced by government policy to choose a wooden ship, regardless of his own preference.

In spite of the fact that Scottish shipbuilders along the River Clyde had been successfully building riveted iron and/or steel ships for more than twenty years, British authorities as late as 1860 adamantly refused to permit the Royal Mails to be transported aboard anything but wooden hulls, and would not approve mail contracts involving the use of iron or steel ships.[10]

---

[10] Alexander Stephens and Sons, Ltd., page number not recorded.

Chapter VIII

The Coal Parade to Panama

A passing vessel reported speaking the *Struan* on July 20, 1882 in latitude 14° N, longitude 27° W (some 2,500 miles into her voyage) southwest of the Cape Verde Islands off the west coast of Africa, indicating that the ship had progressed only a little more than one hundred miles per day—averaging only five knots—along its intended course. Winds and currents must have been more cooperative soon after that sighting, which was reported in *Lloyds' Weekly Index* on August 25, for the *Struan* covered the distance from the location of the speaking, around Cape Horn and up the South American coast to Taboga Island at Panama—some 12,000 miles—in a few days more than three months, arriving October 28, 1882.

Some confusion regarding the actual course and location of the *Struan* arose from the publication in the December 8, 1882 edition of *Lloyds' Weekly Index* of a notation that the vessel had arrived at Tobago[1] on October 24. The similarity in spelling of the two islands may have precipitated the error, but anyone interested in the ship or its cargo could have learned the actual port of arrival by a simple telegraphic inquiry to authorities at either location.

On arrival at Taboga Island (about ten miles south of Panama City out in the Bay on the Pacific side of the Isthmus), Captain Cooper found that unloading would be delayed because of a lineup of vessels bringing coal and other supplies for the railroad locomotives, excavating machines and other equipment being used by the French canal builders. Actual digging on the project had begun January 20, 1882, and most shipmasters at the site were convinced that more could have been done to expedite and organize the handling of incoming cargo

[1] A 25-mile long island off the north coast of South America, in latitude 11° 15' N, 60° 45' W, now associated politically with Trinidad.

to facilitate quicker unloading of vessels. While they waited at anchor in the tropical climate, ships' officers concerned themselves that the accumulation of marine growth on all the hulls below the waterlines would seriously reduce their speed during the next leg of the trip.

Although "irregularities" had been reported in handling of deliveries being made by ship on the other side of the Isthmus (at Colon on the Caribbean) such breaches of business ethics had apparently not taken root at Taboga by the time of the *Struan*'s arrival. At Colon, some ships were alleged to have unloaded half their cargoes, been paid for a full shipload, and then ridden at anchor down the coast for a week or so, only to return and be paid again for another full cargo.[2]

Ten days after the *Struan*'s arrival at Panama, the *Daily Star and Herald* acknowledged the vessel's presence in the area, noting that:

> The English ship *Struan*, Captain Cooper, with 2,400 tons of coal for the Pacific Steam Navigation Company, is at present discharging her cargo at Taboga. She came from Newport, Wales and made the voyage out in 126 days.[3]

After finally discharging his cargo, Captain Cooper took on ballast and departed for Burrard's Inlet,[4] British Columbia, Canada. Because that port had not been developed and was not well known among shipping interests at the time, he recorded his destination as Victoria (Vancouver Island) when he sailed north on December 23, 1882, beginning a 4,032-mile voyage.

The *Struan* was spoken February 9, 1883 at 30° N, 129° W, about six hundred miles southwest of San Francisco, indicating that Cooper was staying well offshore to give himself adequate sea room in case he encountered a storm or adverse winds.

Two and one-half weeks later the *Struan* dropped anchor off the Moodyville Lumber mills in Burrard's Inlet. Before the middle of March desertions from the ranks of seamen began, and by early April

---

[2] David McCullough, *The Path Between the Seas—The Creation of the Panama Canal* (New York: Simon and Schuster) p. 164.

[3] The Panama City *Daily Star and Herald*, November 6, 1882.

[4] The eastern arm of the Strait of Georgia extending 23 miles in an easterly direction into Southeastern British Columbia, Canada; forms Vancouver harbor, one of the best natural harbors on the Pacific Coast of North America. The inlet was explored in 1792 by British navigator George Vancouver, and named by him for Sir Harry Burrard, who had served with Vancouver in the West Indies in 1785. (*Encyclopedia Britannica*, 1992, II, 664).

ten had left the ship, including the second mate, John Stevens, who deserted just as loading was completed.

Sewell Prescott Moody bought the Pioneer Mill on the north shore of Burrard's Inlet in 1864 after it had gone bankrupt twice in its first year of operation under previous owners. He revamped and expanded its facilities, to specialize in cutting high quality large dimension timbers from gigantic trees in the surrounding forest, catering to quality-conscious architects throughout the world. An order for 100-foot long timbers as large as 24-inches square was not out of the ordinary. In some cases, loggers had to put in two or three days hard labor to fell trees large enough to satisfy the requirements of some of Moody's customers.

Moody's mill provided masts and spars for merchant ships and naval vessels internationally, and made cedar shingles not only for the rapidly growing company town—Moodyville—surrounding the mill site, but also for the construction work in the city of Vancouver.

"Sew" Moody is reported to have died in a shipwreck off the coast in 1875, but the Moodyville mill continued to supply high quality timber to buyers worldwide for another 12 or 15 years.[5]

Even before confederation of the eastern Canadian provinces in 1867, both coal and lumber were being exported from what is now British Columbia. By the mid-1880s, coal had surpassed gold as the province's most valuable mineral. Across the Georgia Strait on Vancouver Island, the first recorded export of lumber took place on September 25, 1863 from Chemainus, on the lower east coast a few miles south of Nanaimo, where a coaling station had been established to handle the locally mined fuel even prior to 1859.

Loading lumber and timbers in those early days required a great deal of hard work. At Chemainus, a "bull gang" of Chinese laborers moved the timber from the mill yard to the sailing vessel on heavy two-wheeled carts pushed by hand along a series of planks laid over the ground to form a track. If a cart wheel happened to slip off the plank "road" into soft dirt, loading was delayed until the derailed load was cleared from the track or a new series of planks laid.

Fifteen years later, at most mills, teams of horses were doing the heavy work, and more progressive operators had begun to install steam winches utilizing a series of cables to make the loading easier. Exactly which loading methods were in use at the Moodyville mill in 1883 has not been determined.

---

[5] Aynsley Vogel and Dana Wyse, *Vancouver: A History in Photographs* (Banff: Altitude Publishing Co.) 1993, p. 12.

Captain Cooper found some difficulty in signing a new crew. The Canadian Pacific Railway was building its transcontinental line eastward from Port Moody to meet another group forging westward from Montreal, and employment opportunities were plentiful in the bustling new community. Unable to find a suitable licensed new second mate, he hired a 34-year old Falmouth seaman, Joseph A. May, as boatswain (bos'n) for the trip to Melbourne. In order to attract enough competent sailors to take the ship to sea, Cooper found it necessary to agree with many of them that they would be discharged within 48 hours of arrival at the next port of call (Melbourne, Australia) where, he felt certain, they could be easily replaced.

Safely moored in a protected harbor, the *Struan* suffered no damage from high winds and heavy rains that blanketed the area on a number of occasions while at Port Moody, but newspaper notices invited interested persons to an auction to be held April 28 to dispose of the wrecks of two vessels which had been swept onto the rocks while trying to enter Esquimalt Harbor during a raging storm.

The New Westminster *Mainland Guardian* on May 12, 1883 carried information that:

> The ship *Struan*, Captain Cooper, now loaded at the Moodyville Mills with lumber for Australia, will go out tomorrow; there are two other vessels (barques) loading lumber at Moodyville, and one ship, the *Lady Bowen*, at Hastings.

The voyage from Moodyville (now part of the Vancouver, British Columbia harbor) to Melbourne, Australia—very close to 7,500 nautical miles—required about 110 days. The *Struan* dropped anchor in Port Phillip Bay off Melbourne on September 2, 1883, and Captain Cooper deposited his ship's articles at the Merchant Marine Office the next day.

The *Melbourne Age* announced the vessel's presence on September 3rd:

> Arrived, *Struan*, ship, from Esquimalt Bay (Burrard's Inlet) [with lumber], 13 May[6] Couche, Calder and Co., agents.

Captain Cooper hired a 21-year old second mate, John McCleery, at Melbourne, but the circumstances of his arrival aboard the *Struan*

---

[6] This date indicates the day of departure on the present voyage.

are somewhat clouded. The ship's papers show plainly in ink entries that he was engaged August 27, 1883, some six or seven days *before* the *Struan*'s arrival at Melbourne, but there are some illegible notes on the ship's articles regarding his engagement, apparently written in pencil, that might have shed some light on the mystery of when he actually came aboard, had we been able to read them.

A major topic of conversation among members of the seafaring fraternity while the *Struan* was in Melbourne involved the introduction of a new periodical called the *Australian Illustrated News* with a format similar to the well-established *London Illustrated News* in England. Many readers thought that the new Australian magazine was equal to its home-country predecessor publication both in form and content. Some critics were particularly impressed with the quality of its artwork. Especially noted was a rendering of a shipwrecked sailing vessel impaled on a rocky ledge along the Australian coast. The pen-and-ink sketch was considered by many comparable to a photograph in detail and quality.

The captain's hope for a quick unloading of his vessel again turned out to be wishful thinking. Two months elapsed before the lumber and timbers from Moodyville were all removed from the *Struan* and ballast was taken aboard for the relatively short trip to Newcastle, N.S.W. Sixty days was the average time required for both unloading *and* reloading other types of cargo such as coal, grain, rice etc.

The extra time required for the handling of timber cargoes was one of the major reasons that many shipowners, if given a choice, only accepted lumber if the freight rate were high enough to cover the costs incurred by the extra time in port.

Freight rates being quoted at that time in the Puget Sound area for the transportation of lumber and heavy timbers to Pisagua and Iquique, Chile, on the west coast of South America, were in the neighborhood of 42 to 43 British shillings, about U.S. $10.50 per thousand board feet,[7] almost exactly the same price that was paid for a thousand board feet of lumber at the mill. For the voyage to Melbourne, perhaps 1,600 nautical miles longer, the rate can be estimated at about U.S. $13.00, making a one-and-one-quarter million foot cargo[8] worth about $16,000 in transportation fees. Considering

---

7 A board foot of lumber is the equivalent of a one-inch thick board 12 inches wide and one foot long. Thus, a one-inch thick board 6 inches wide and 10 feet long would contain 5 board feet, while a 2 x 8 inch plank 12 feet in length would measure 16 board feet (fbm).

8 On her final voyage in 1890, the *Struan* carried almost 1,500,000 board feet of lumber and timbers. It is likely, however, that Captain Cooper chose to load a somewhat smaller cargo.

that two and one-half months were required for loading, four months in transit, and two more months for unloading—almost three-quarters of a year in total, most shipowners would not have considered the voyage particularly profitable.

The inward list of members of the *Struan* crew prepared for submission to the Newcastle port authorities by Robert B. Wallace, the ship's agent, on November 13, 1883 totalled 23 names. One of these, Peter Thompson of Dublin, must have been included in error, since the ship's Articles reveal that although he was hired at Newport for the voyage, he had not appeared by sailing time and was listed as "never joined."[9]

Twenty-two hands all told, including the captain, officers and crew seems to have been the normal personnel complement aboard the *Struan* under British ownership. Several surviving crew agreements made over an eight-year period specify that number, always noting that "any additional are extra." The frugal Norwegians, however, when they purchased the vessel in late 1889, operated the ship with a total crew of only 18, and made one 40-day trip in ballast from the west coast of South America to Port Townsend, Washington with only 17 on board.

Included in the four ships and 12 barks loading at Newcastle when the *Struan* arrived was the ship *Rock Terrace*, Captain Hutchinson, owned and operated by Troop and Son of Saint John, New Brunswick, and named for the Troop estate there, loading coal for Wilmington via San Francisco. One of the senior captains serving in the Troop fleet, Andrew Freeman Kenney had been the first commander of the *Rock Terrace* when it was launched at Saint John in 1875 by shipbuilder David Lynch. Kenney would later be skipper of the *Tamar E. Marshall* when it rescued the crew of the dismasted *Struan* off the North Pacific coast of the United States in December 1890.

Captain Cooper disposed of the *Struan*'s ballast as directed by the port authorities and berthed his vessel at Dyke Wharf to await her turn to load Wallsend coal for Panama. That opportunity came on November 25, 1883 when she went alongside hydraulic crane No. 6 to take on a full load of 2,250 tons.

There were at least a dozen operating collieries in the Newcastle area at the time, and information furnished by the seven of those

---

[9] Under British law at the time, a captain could be subjected to a nominal fine (10 shillings, or less than U.S. $2.50) for each unexplained instance in which a person, other than a passenger or rescued shipwreck survivor, was carried aboard a vessel in the foreign trade unless he or she was a properly engaged member of the ship's crew.

which made weekly reports for publication in the *Newcastle Morning Herald* disclosed that a total of 30,148 tons of coal had been shipped during the week ending November 10, 1883 to twenty different ports in various parts of the world.

Five firms with banking offices in Newcastle, placed ads on the marine news pages of the *Herald* to attract shipmasters who might need bank services to arrange payments for cargo, etc. Provisions needed for the outbound voyage were available to ships at various waterfront locations. The day after the *Struan* arrived, potatoes were quoted at £2 10s (about $12.50) per *ton*; hams and bacon at the equivalent of 20¢ per pound, butter at 15¢ per pound and eggs at 35¢ per dozen.

One feature not often found in newspapers of that time appeared in the marine section of the *Newcastle Herald*'s December 4 edition while the *Struan* was being loaded. It was an apologetic correction of a mistaken departure date for a vessel from a smaller port near Melbourne in the Australian state of Victoria carried in the previous day's notation. By mistakenly adding four days to his passage time (on a seven hundred-odd mile trip) to Newcastle, the *Herald* realized that it had left the captain open to ridicule by his contemporaries or to criticism by the vessel's owners:

> In our yesterday's issue, the date of departure from Geelong of the barque *Frederica* was given as 22 ult. This should have read the 26th—a very material difference.

The *Struan* had been cleared and was ready to leave Newcastle December 6, but did not actually depart for Panama until December 9, 1883.

Less than two weeks after his arrival in Panama, Captain Cooper was concerned by a dispatch published in the local *Daily Star and Herald* (April 25, 1884) relating that a severe earthquake had been experienced throughout the eastern counties of England, during which the Colchester, Essex, church spire was broken and other serious damage done. Having been born and raised in the Essex area, Cooper naturally would have been apprehensive about his family's safety and possible damage to their property. It is likely that he a telegraphed asking for assurance of their well-being. For the convenience of the maritime fraternity and passengers aboard ships in the harbor who could not receive messages on a daily basis, the *Star and Herald* included in its regular editions almost every other day, both in

THE English ship *Struan*, Captain Cooper, arrived in the bay on the 13th instant with 2,250 tons of coal for the Pacific Mail Steamship Company. She is from Newcastle, New South Wales, and has taken 140 days to make the voyage, during 60 of which she was within 500 miles of this port. Notwithstanding the length of the voyage her officers and crew have arrived in the best of health.

Verbatim reconstruction of a news item from the Panama City *Daily Star and Herald*, April 16, 1884.

English and Spanish, a list of people for whom messages were being held at the offices of the West India and Panama Telegraph Company.

Because steam tugs in the Panama Bay and anchorage area were few and obviously overworked, the lighterage fees for ferrying cargo from ship to shore, or vice versa, were $1.25 per ton. Dock space being severely limited, the 2,250 tons of coal brought from Newcastle, N.S.W. would have cost over $2,800 to unload if the cargo had been lightered ashore. Since the ship's agent at Panama had been able to negotiate a sale of all the coal to the Pacific Mail Steamship Company during the *Struan*'s transit from Australia, it is likely that the purchasers arranged to have their property transferred directly to a floating coal barge or coal hulk serving as a supply station or depot for their steam-powered ships.

Proprietor James Boyd of the *Daily Star and Herald* let little that occurred in Panama City escape his journalistic eye. He criticized the police for their failure to stop nighttime burglaries and street robberies of pedestrians; he attacked the municipal government for allowing garbage to accumulate on the streets, and when the large two-story wooden Valparaiso Hotel in Panama City caught fire on Sunday evening April 27, resulting in a conflagration that ultimately consumed 38 buildings in a four-acre area, Boyd complained that gunpowder, coal oil and liquor were stored in the same building where people slept. (He reported that the captain of the steamer *Casma* saw the reflection of the fire while 46 miles at sea.) Boyd harangued the authorities for not adequately protecting property in the burned district from riotous looters during the emergency, and he summed up his comments regarding the catastrophe and the government's actions or lack thereof, as—"No water, no firemen, no system."

In early June, Boyd wrote the first of a three- or four-part editorial about the city's slaughter house, alleging that dead animals were being delivered to the facility at night and being marketed as freshly killed. When the official in charge of the "matadero" failed to respond promptly to questions posed in his first editorial, he followed with explicit charges brought by various witnesses, with the result that Captain Cooper and other shipmasters declined to obtain fresh meat from Panamanian sources for their outward voyages. Vessels were sailing regularly using the "substitutes at the master's option" endorsement in the ships' logs with explanations of the circumstances.

Cooper was understandably anxious to make his departure from Panama as quickly as possible after unloading and ballasting had been completed. As it turned out, his choice of departure time could hardly have been worse. On the Friday immediately preceding his planned June 7, 1884 sailing, about an hour after noon,

> a waterspout was clearly seen in the Bay. It resembled an immense corkscrew, and after a few minutes its lower part was taken up and disappeared in the clouds. This remarkable meteorological condition is seldom observed here.[10]

The same storm that blew down the dance hall at Panama City the next day caught the *Struan* at sea, causing the loss of two small spars.

---

10 Panama City *Star and Herald*, June 7, 1884.

The *Star and Herald* reported the incident to its readers, details of which were telegraphed to London.

> The English ship *Struan*, Captain Cooper, which left the Bay in ballast on Saturday last for Valparaiso, was struck by a squall in the Gulf; two small sticks were carried away. She anchored off Pacora, and sent a boat to Panama [about twenty miles], replaced them, and continued on her voyage on Tuesday.[11]

After 45 days becalmed off Buenaventura, Colombia (only about one hundred miles into her voyage) she returned again to Panama October 2, to clean her bottom and reprovision. Captain Cooper notified the owners that after many days of ineffectual struggle against the "doldrums"[12] he had not been able to proceed farther south than the latitude of Manta, Ecuador (1° south of the equator).

From previous voyages to ports on the west coast of South America, Cooper was aware of the existing northerly current (the Humboldt current) all along the Chilean coastline—in some cases so strong that vessels heading north after rounding Cape Horn were carried past their intended destinations and had to circle back to reach their scheduled ports of call from the south.

Although there are no supporting records, it is likely that Cooper suggested to the owners that the voyage to Valparaiso (for which he had no cargo and was sailing in ballast) be abandoned and that the *Struan* proceed instead to San Francisco, where needed repairs to the vessel could be made and a wheat cargo obtained for the homeward voyage. His ideas having apparently been approved by the managing owner, he left Panama for San Francisco on October 30, 1884.

---

11 Panama City *Star and Herald*, June 13, 1884.

12 Those parts of the ocean near the equator that abound in unpredictable squalls, calms and light baffling breezes, occurring in no predictable sequence, making progress through the area difficult or impossible for a wind-driven vessel. Squalls are generally not of great duration, but calms may last for long periods of time.

Chapter IX

Admeasurement, Tonnage and Cargo; Carpenter's Measure

Some Scottish shipbuilders in the late 18th century had developed and were using a formula for the measurement of their own vessels that they called "carpenter's measure," a simple, easily understood and readily calculated system that essentially called for the length of a vessel to be multiplied by its breadth and that product by half of its breadth again, because the formula allowed the builder to assume, regardless of the actual depth, that all vessels would be one-half as deep as they were wide. The product of these three dimensions was to be divided by 94 to arrive at the vessel's tonnage. Apparently no adjustments were made for spaces occupied as crew quarters or for other areas not available for stowage of cargo, nor for differences in hull shape such as blunt or sharp bow lines or flared hull patterns designed to increase speed, etc.

With the adoption of the first Registry Act in 1786, Parliament prescribed for the measurement of ship's tonnage a method almost identical to the Scottish one, which should thereafter apply to all classes of vessels. Those instructions and requirements continued in force until the end of 1835.

Computation of the *Struan*'s tonnage under those regulations would have been as follows:

211.80	(length in feet)
x 39.30	(breadth in feet)
8,323.74	
x 19.65	(assumed depth as 1/2 of breadth)
163,561.50	
÷ 94.00	
= 1,740.00	tons

Early in my research, I discovered a method closely related to "carpenter's measure" in an old encyclopedia volume, though there was no explanation in that record of when and where it was developed, what it was called or which countries or maritime associations favored its use. Based solely on the fact that vessel tonnage calculated by this variation was substantially greater (see example below) than if computed by the original "carpenter's measure," it seems safe to conclude that this variation must have been in use *prior* to 1836 if it is of British origin.

The paraphrased instructions are as follows:

Deduct from the overall length in feet three quarters of the breadth (width or beam); multiply this result by the breadth (width or beam) itself, and that product by the depth; divide the resultant answer by 95 and the quotient thus obtained is the register tonnage for payment of fees.[1]

By this variation of "carpenter's measure," the *Struan*'s tonnage would have been calculated as follows:

```
    211.8  (length in feet)
   – 29.5  (3/4 of breadth)
    182.3
   x 39.3  (breadth)
  7,164.4
   x 24.2  (depth)
173,378.5
   ÷ 95.0
  = 1,825  tons
```

Throughout Great Britain's modern history, particularly after the beginning of the 19th century, there was a recurring demand by shipbuilders, shipowners and the entire maritime community for improvement in the methods used to measure the tonnage of vessels.

Realizing that the rules then in effect unfairly inflated the tonnage measurement of some vessels beyond their actual capacity, interested parties in 1833 persuaded the authorities to appoint a Commission on Tonnage to examine the situation and recommend changes. This action ultimately resulted in Parliamentary passage and adoption on January 1, 1836 of what was then identified as "New Measurement."

---

[1] Funk and Wagnall's *New Standard Encyclopedia*, 1934.

This later law required that vessels henceforth be measured not in feet and inches but in feet and tenths of a foot, ostensibly to facilitate a more accurate and less convoluted computation. The calculation, though considered by some "an easy arithmetic process," was admittedly "somewhat more complex" than past systems for determining the same measurement, especially before mechanical or electronic calculators, when repetitive multiplication and division of six or seven digit figures required a substantial expenditure of time and paper.

The method for determining the "New Measurement" required that a ship's length of deck, from stem to stern, be divided into six equal parts, and that two horizontal cross-sectional or athwartship measurements be taken at the first, third, and fifth points of division—one such measurement to be made about midway between the weather deck and the main deck in the hold, and the other one about halfway between the main deck and the top of the keel. Only one measurement of depth was required at each of the three cross-sectional stations.

All of these measurements were then utilized (some two or three times) in a somewhat complicated formula. The details are not presented here, but a complete account and specific directions may be found in an analysis by a naval architect writing on the subject in 1852.[2]

Criticism of the system mounted when the possibility of easy evasion of the regulations was exposed in print. Skimpy explanations and lack of specific instructions given to surveyors by the authorities enabled individuals to interpret the act in whatever manner best suited their private purposes.

One presumably unintended and initially unforeseen effect of this act was to alert shipbuilders to the fact that a longer but not so deep hull permitted the same or even greater cargo to be carried, but actually reduced the registered tonnage figure computed by this method. It seems entirely possible (though not proven) that this act of Parliament may have done more to encourage the building of larger, especially longer, ships than any other factor.

The next revision of the Tonnage and Registry Act, promulgated in 1854, required almost five years for completion by a Commission consisting of shipbuilders, shipowners and naval architects. When members of the Commission could not agree on any of the proposals

---

[2] George Moorsom, *A Brief Review and Analyses of Laws for Admeasurement of Tonnage*, London, 1852.

before them, they looked at a scheme advanced by naval architect George Moorsom, who theorized that measurements must be made of the *internal* dimensions of a vessel to make the system fair in its application both to thin-skinned iron ships that were beginning to appear, as well as to the heavy-hulled timber vessels still popular with many shipowners at the time.

Though it was perceived as more equitable in its application to the broad range of new hull shapes and ship designs, this system required, depending on a vessel's length, that cross-sectional measurements be taken at as many as 12 locations, and it entailed substantially complex computations to produce a tonnage figure.[3]

This 1854 system of measurement was the one under which the *Struan* was surveyed and registered upon launching at Saint John, New Brunswick in 1877.

The *Struan*'s original registration papers, dated May 29, 1877, reflect the results of the detailed computation:

Underdeck tonnage	1,398.24
Plus lazarette and wings	18.70
Plus forward roundhouse (less galley)	43.57
Plus roundhouse aft	49.25
	1,509.76
Deduct allowance for crew space	36.47
Register Tonnage	1,473.29

It can readily be seen by comparing the *Struan*'s registered tonnage figures, calculated under the regulations then in effect in the 19th century, that the earlier computations had actually resulted in substantially higher vessel tonnages. Since many vessels were contracted for construction at that time on a basis of so many dollars per registered ton equipped for sea, there is no doubt that some shipbuilders viewed the 1854 changes as detrimental to their interests.

Shipowners and operators, on the other hand, were required to pay various port fees and other charges based on registered tonnage wherever their vessels called to load or unload cargo. They were, therefore, happy with the 1854 change.

This diversity of opinion between interested parties may have contributed to the lengthy delay in acceptance by Parliament of the 1836 and 1854 changes as well as the failure of a previous attempt (in

[3] MacGregor, p. 168.

1831) to reach any agreement.

It should be noted that registered tonnage does not refer to the weight of cargo that can be carried, but is rather a calculation of capacity by volume. In this connection, one hundred cubic feet comprise a measurement ton. It is therefore incorrect to assume that a vessel was overloaded merely because it carried a cargo weighing more than its register tonnage.

There are many records of unbelievably heavy cargoes being taken aboard by sailing ships. Wallace mentions the ship *Canada* built in 1891 at Nova Scotia's Kingsport yard, well after adoption of the Plimsoll Mark legislation, (see p. 41) registered at 2,137 tons and alleged to have actually carried 3,600 tons.[4]

The register tonnage of the *Guiana*, referred to in the chapter on insurance, was 1,265, which translates into 126,500 cubic feet of below-deck or shelter-deck space available for the cargo. On her voyage from the Philippines in 1885-86 she carried 1,929 gross tons, or 4,320,960 pounds of sacked sugar. Since that commodity weighs about a hundred pounds to the cubic foot, such a load would have required only about a third of the available cargo space on the *Guiana*, though the loading of additional cargo would have taken her below her Plimsoll mark.

43,210	cubic feet needed
÷ 126,500	cubic feet available
= .34	of hold space needed for sugar cargo

In the case of the *Clythia*, discussed in the chapter entitled "The Clythia Marbles," her cargo of rough marble blocks is estimated to have weighed approximately 172 pounds per cubic foot, needing less that three-fifths of the space occupied by sugar of similar weight. This explains why the *Clythia's* captain welcomed a partial cargo of crated furniture to utilize some of the empty space in his vessel's hold not needed for the heavy marble, at a higher freight rate than the marble demanded.

---

[4] Frederick William Wallace, *Wooden Ships and Iron Men* (Charles E. Lauriat Co.) 1937, p. 316.

To have carried 3,600 tons, the cargo must have (1) weighed at least 36 pounds per cubic foot (about equal to some species of freshly cut fir timber), been tightly stowed, perhaps including a deck load; or (2) been a commodity weighing in excess of 40 pounds per cubic foot (like partially dried oak timbers).

$$\begin{array}{r} 100 \\ \div\ \underline{172} \\ = \underline{.58} \end{array}\text{, say .6 or 3/5ths}$$

$$\begin{array}{r} 3/5 \\ \underline{\text{x } 1/3} \\ = \underline{3/15} \end{array}\text{, or 1/5 of hold space needed for marble cargo}$$

At the other extreme would be a shipment involving a relatively light commodity.

The largest cargo of cotton ever loaded from Norfolk, Virginia up to that time was put aboard the full-rigger *William Law*, 1,599 registered tons, in early 1883—6,723 bales totalling more than 3,120,000 pounds—about 1,393 gross tons load, substantially less than the vessel's registered tonnage as a result of cotton's relatively light weight.[5]

The register tonnage and cargo weight of the *Struan* on her final voyage will be discussed in the chapter relating details of that passage.

Although the weights of various types of cargo were important to shipmasters and owners in the *Struan*'s time, the peculiarities, hazards and physical properties of certain cargo items were even more critically considered.

Captains and their crews were more fearful of the potential for fire at sea than they were of any other situation that could face them on a long trip. Coal cargoes, for instance, were particularly problematic because of the possibility of spontaneous combustion. Although it was quite well known that coal could, and frequently did, catch fire aboard a vessel in transit without any apparent reason, the causes were not well understood.

Some captains refused to load wet coal while others believed that the dust-like particles present in mill-run coal were the problem. As it became more common in the late 19th century to screen for size and wash coal as it came from the mines, some of these potential concerns were eliminated.

Probably the best known record of a coal fire at sea aboard a sailing vessel is that involving the Troop and Son bark *James W. Elwell* of Saint John, New Brunswick in 1872. Captain John Wren of Saint Andrews, New Brunswick lived to relate the story of a 72-day ordeal in an open boat when he and 12 of the crew abandoned the *Elwell* six

---

[5] *Ibid*, p. 232.

days after they first discovered tattletale wisps of smoke rising from a ventilator shaft while changing course in a wear-ship maneuver 160 miles off the Patagonian coast of Chile.

Repeated storms, high winds, chilling cold, lack of shelter and impending starvation gradually whittled the 13 aboard to only three—the captain, the stewardess who had been the wife of the ship's cook that died prior to discovery of the fire, and one sailor. These three survivors were eventually picked up by the steamship *Tropic*, which had been proceeding on a route different from the one usually followed. But for that variation in course, Captain Wren insisted, he and the other two would also have perished.[6]

In one two-week period in March 1882, the ships *Novara* and *Norval*, both carrying coal from different ports of loading to separate destinations were burned near the equator in the Pacific Ocean west of the Galapagos Islands, within a few hundred miles of each other. Crews of each vessel escaped the fire in their boats, all of which were picked up by passing vessels.[7]

A marine reporter commenting on the coal trade in late 1884 argued that the San Francisco customs authorities should permit coal carriers to leave their hatches open upon arrival in the Bay, since "there is no question but that certain cargoes have become heated through lack of ventilation after arrival." He also noted that some importers had experienced difficulty arranging transportation for Scotch Splint, West Hartley and Higginson's Brymbo coals because insurance companies were demanding double last year's premium rates on the ground that these varieties were very easily combustible and therefore unsuitable for long voyages.[8]

Shipmasters were always on the alert when loading drums of petroleum products. There was a flourishing trade in transporting this commodity from Philadelphia and other U.S. east coast ports to Europe in the *Struan*'s time. An officer carefully inspected every container as it came aboard to be absolutely certain that no "leakers" were shipped.

One hazard not even suspected previously in the handling of petroleum products contributed to the loss of the ten-year old ship *Regina*, Captain Joseph Bain, in November 1883. Enroute from Philadelphia

---

[6] *Ibid.*, pp. 239-247.

[7] Murray Lawson, *Record of Yarmouth Shipping*. Publisher and date of publication not available. A copy, for reference use only, is available at the Halifax, Nova Scotia library.

[8] The San Francisco *Merchant*, December 19, 1884, p. 75.

to London with drums of petroleum, the vessel encountered heavy weather and had the main hatch cover damaged by the tremendous seas. After several attempts at putting a spare sail over the leaking hatch almost cost the lives of those attempting the task, the effort was abandoned, and the wallowing vessel continued to take the seas through the damaged cover. As water filled the hold, the drums of petroleum in the lower cargo space broke loose and floated up under the deck with such force that the decking and some of the deck beams were dislodged; the vessel was pounded to pieces by very heavy seas on the outside and the projectile-like drums of oil from the inside. At least a half dozen of the crew were lost before the survivors were picked up by the bark *Helen Finlayson* of Ardrossan, a seaport and shipbuilding center in Ayrshire, southwest Scotland, several days after the hull disintegrated.[9]

Even such a commodity as wheat, generally considered a relatively safe cargo, under certain circumstances has been known to cause the death of seafarers. In 1878 the bark *Viking* of Yarmouth, Nova Scotia, outbound from Philadelphia with a cargo of wheat, encountered heavy weather and sprang a leak. Before long the pumps became clogged with grain, and the captain himself went down the hatch serving the well-pump in an attempt to clear the stoppage. At the bottom of the shaft the gas generated by the water-soaked and rotting grain was so powerful that the shipmaster was overcome and fell unconscious into four feet of water. A seaman who saw his predicament and went to his assistance met the same fate. The second mate was able to take down a rope and recover the bodies, but the doomed vessel had to be abandoned only a few hours later. It is reported that crew members were taken aboard a Spanish brig.[10] [11]

One of the most difficult and time-consuming cargoes to load aboard sailing ships was baled wool. Being even lighter than cotton similarly packed, it always provided a challenge to the loading crews. In order that as many bales as possible could be carried, the wool was said to be "screwed" into position in the hold of a vessel with the aid of jacks, levers, shores, planks and whatever other mechanical devices might be used to force the bales into the smallest possible space.

From about the 1860s, rough bales of "grease" wool from the

---

[9] Wallace, pp. 201-204.

[10] *Ibid.* pp. 229-230.

[11] It is probable that expansion of the water-soaked grain may have put sufficient pressure on the hull planking to make the leak worse, eventually contributing to the loss of the vessel.

Australian outback sheep stations were hydraulically compressed at dockside just prior to loading and then put aboard ship as soon as possible lest the straps be broken by the terrific pressure of the wool to expand, especially if exposed to sunlight for any length of time.

A vessel carrying wool alone could never be loaded heavily enough in the lower hold to guarantee stability in a heavy sea. Some heavier cargo or ballast was required to be in place before the wool was put aboard.

Even after the difficulty and delay in loading and packing a wool cargo, a shipmaster had to hold his breath all the way back to England for fear that some of the bales might have gotten wet while in transit to the dock. Any dampness or moisture trapped inside the bales could bring on the dreaded fire from spontaneous combustion. More than a few well-known vessels met their fate in that manner.

Several were destroyed in dockside fires while loading or unloading wool. The 1865 discovery of smoke aboard the *Fiery Star* 18 days out of Brisbane for England is perhaps the most harrowing tale, involving a shortage of lifeboats for passengers and crew, a capable and courageous first officer, and the last minute rescue of those still aboard fighting the fire three weeks after it was discovered.[12]

In some instances, vessels were abandoned at sea under conditions that appeared to justify such action at the time, though later developments may have called into question the wisdom of the decision to abandon ship.

The Monday, July 22, 1889 issue of *Shipping Gazette and Lloyds' List* carried the story of the Italian bark *Vincenzo Perratta*, abandoned September 18th, 1887 in approximate latitude 36° N, longitude 54° W, about seven hundred miles south of Newfoundland. One year, six months and 16 days later, the vessel was sighted, still afloat, having drifted some 1,400 miles in a general southwesterly direction to the vicinity of Watley's Island in the Bahamas. During that time, reports on the location of the derelict were filed with authorities 27 times, the most recent reporting that the mizzenmast and about ten feet of the mainmast were still standing, the foremast gone, the jib-boom broken off and the port anchor still in position at the bow.

---

12 Basil Lubbock, *The Colonial Clippers*, (James Brown and Son, Publishers) Glasgow, 1921, pp. 118-119.

Chapter X

Destination: Admiralty Court

Captain James Cooper brought the *Struan* into San Francisco Bay on January 2, 1885, 64 days from Panama in ballast, with the vessel consigned to J.F. Chapman & Co., shipping and commission merchants, 22 California Street. The anchor had hardly hit the bottom before word spread that the ship had experienced a strange occurrence on her trip, and reporters from most of the San Francisco dailies pressed the captain for particulars of the voyage. "We left Panama October 30 with a light breeze from the NW," he told them,

> and passed Cape Maylo on the following day; from there we experienced light baffling winds from W to SW, which delayed us on the coast for three weeks. Then we got the wind from the southward; between 100° and 108° west longitude had light winds from the eastward, freshening into NE trades which carried us to 20°N latitude. Between there and 25°N had light variable winds and calms which lasted nine days. Then the wind came fresh from the SW, with thick weather, veering to the NW with strong gales and squally weather which continued up the Farallone Island light.

Sensing the growing impatience of some of his listeners, Cooper continued with that part of the story they had really come to hear:

> On December 6, in latitude 4°17'N, at 11:30 PM we felt two distinct earthquake shocks, the first lasting about 30 seconds, with an interval of 20 seconds; the other lasting about 15 seconds, both shaking the ship as though she was passing over a reef. The sea at the time was very smooth with a light breeze from the SW.[1]

---

[1] The San Francisco *Evening Bulletin*, January 3, 1885, p. 3.

At the British consulate on January 3, 1885 Cooper paid off about half his crew, including eight men who had shipped at Panama on a "by the run" basis, with the understanding that they would be released within 48 hours of arrival in San Francisco. Each received £7 for the 64-day trip, less than $35 in U.S. money, or about 50¢ per day for the voyage. Vice consul Christian Mason made an indorsement on page ten of the ship's articles in which he wrote:

> I certify that the within named Wm. Crowley, A.E. Weaver, R.P. Rasmussen, Torvold Anderson, James Simmons, D. McMillan, Tens Cronin, John Meyrick, A Mutten, and Robert Ribbons have been discharged at this port with my sanction; Crowley and Weaver by mutual agreement, and the others through their terms of service having expired; that their balance of wages as placed against their respective names has been paid to them in cash and their effects delivered to them.

Remaining on board the *Struan*, then, during her time in San Francisco and for the trip to Grimsby loaded with wheat, were Captain James Cooper, first mate Samuel Lemon, Norwegian carpenter Neil Christian, and two ABs from the original crew hired in Newport at the beginning of the voyage—Swedish sailor Andre Gutlafson and William Shaw. Also remaining were four ABs taken on at Newcastle, NSW—Samuel Donnelly of Belfast, Britisher C. Dummett, Scotsman John Gilbert, Nova Scotian John McLeod, and steward Anthony White, engaged at the same port. Rounding out the list of stayaboards was AB William Hodgson, who had sailed as a boy and was rehired as an able seaman upon the expiration of his indenture as an apprentice, and ship's boy Fred Daniels. The originally employed second mate on the vessel, John Stevens, had deserted at Burrard's Inlet, and John McCleery, a replacement officer taken on in August 1883 at Melbourne lasted less than 14 months and was discharged "by mutual consent" at Panama. There is no record that Captain Cooper ever hired another second mate, even upon departure from San Francisco. Examination of the crew list also leaves open to question the matter of who did the cooking during the three months the *Struan* was in port at San Francisco and on the trip home, since the oriental cook from Whampoa, China, who had been

---

[2] The San Francisco *Morning Call*, Monday, March 9, 1885, p. 4.
[3] The San Francisco *Morning Call*, Thursday, March 26, 1885, front page.

hired in Melbourne, was discharged in Panama, again "by mutual consent," and no replacement ever showed up on the ship's roster.

All shipmasters arriving on the Bay in early 1885 were being warned by their colleagues about the need to exercise extreme vigilance to protect their vessels against what the newspapers called "thieving wharfrats." "Not content with stealing grain from the seawall, and anchors and chain from the lighters, they now devote their attention to making away with boats and skiffs which are left by their owners in an exposed position. If the boats happen to be well drawn up on the davits, they do not scruple to cut the ropes and let them drop into the water."[2] One waterfront policeman arrested two men caught in the act of "cutting up a hawser coiled on deck into pieces they could carry conveniently."[3] With more than half a full crew aboard, Cooper had no trouble arranging a 24-hour watch.

The *Struan* reached the city by the Golden Gate almost in a class by itself. Although one of about three dozen vessels arriving that week with the intention of taking out wheat cargoes, she was the only ship that had been chartered prior to coming into the Bay—to load grain "on or about March 30." Most arriving vessels at that time showed up without prior cargo commitments, making it necessary for captains and ship's agents to negotiate for loads upon terms in effect at the time. The *Struan*, then, had a distinct advantage over other ships. The warehouses were filled with wheat, and the owners and the grain brokers were anxious to ship it, but almost every passing day saw more vessels being sent out into the stream in ballast. Rates for transportation of grain from San Francisco to major British ports had been reduced some 35% since the *Struan*'s previous loading of a wheat cargo there. The rate of more than three British pounds per ton in 1881 had fallen to a bare two pounds per ton, and some shipowners chose to take their vessels out of service in an attempt either to force the rates up or wait out the existing low charges.[4] Ships removed from service were listed in the marine sections of the daily papers as "disengaged," and on some days the list of such vessels almost equalled the number of "engaged" vessels actively seeking cargo.

Export values of wheat in cargoes cleared by customs authorities from San Francisco in early 1885 were about $1.30 per cental[5] against the $2.03 figure that had prevailed in 1881. A large number of lots of

---

4 The San Francisco *Morning Call*, Saturday, February 28, 1885.

5 A name for a weight of one hundred pounds of a commodity, especially grain, suggested in 1859 by a lawyer named Danson, but not formally legalized for use in trade by the British government until about 1880.

wheat were held by brokers and others under orders from owners in the interior for sale at $1.32 or $1.35 per cental or better, since money was easily available to carry the grain at around six percent and warehouses were cutting storage rates to accommodate shippers, to below 75¢ per ton in some cases.

During their stay in San Francisco, officers and crew of the *Struan* were surprised to find, in the daily press, a detailed map of Burrard's Inlet, British Columbia, showing the site of the proposed city of Vancouver, from which they had departed earlier in the same voyage with lumber for Melbourne, Australia.

Cooper's major concern was the poor condition of his vessel and the pressing need to arrange for major repairs in time to be ready for the looming March 30 loading date. He sent a telegraphic message to the managing owner in England, explaining the need for a major overhaul of the ship and requesting that funds be transferred to his control for that purpose. He was well aware, from the sluggish manner in which the vessel answered its rudder on the voyage from Panama, that the *Struan*'s bottom was badly fouled by its lengthy stay in tropical waters. He was also apprehensive that temporary repairs made to storm-damaged items after loss of at least one "stick"[6] off Panama needed to be made permanent. It is certain that he found no humor in an article in the *Along the Wharves* column in one of the daily papers detailing the removal of *four tons* of barnacles and marine growth from the hull of another similar-sized vessel just a few days prior to his arrival in port.

Having dispatched his report and petition for financial resources to the owners' representative, the captain set about, with the ship's agent, arranging drydock space for the required overhaul of his vessel. He also haunted the telegraph office for word from Liverpool. When he had received no answer after a few days, he is believed to have

---

[6] A telegraphic report to Lloyds of London on June 7, 1884 reported the *Struan* had been struck by a squall in the Gulf of Panama and "two sticks were carried away." Presumably, the "sticks" referred to spars, not masts, since the vessel was able to effect repairs without dockside help.

[7] This date is in conflict with the March 15, 1885 date entered on papers filed in the bottomry action before the Admiralty Division of the High Court of Justice when the *Struan* reached Great Grimsby. It is highly unlikely that the firm furnishing funds under the bond would have guaranteed payment of drydocking and repair fees prior to Cooper's signing the papers, or that he could have had the repairs made without such a guarantee if he did not actually have the cash in hand. It is more probable that the date of initiation of the bottomry bond was about January 15, since the *Struan* emerged from the drydock about February 25 with all repairs completed.

directed an even more pressing wire to the vessel's owners. No records exist of the telegraphic exchanges involved, but after about ten days in San Francisco Captain Cooper finally realized that no funds were to be forthcoming from the *Struan*'s owners, and that any action to put his ship into condition for the long voyage back to England would result entirely from his own efforts. Whether the managing owner told him directly that funds could not or would not be sent, or whether Cooper was left hanging by an absence of any reply to his messages is not known.

Under the circumstances, the captain had only one alternative. He visited shipbrokers and maritime banking firms, and on January 15, 1885[7] signed a bottomry bond, essentially a mortgage on the ship itself, which would have to be paid off within five days of the arrival of the *Struan* at Great Grimsby. Under then-existing law, captains had authority to take such action without either the permission or knowledge of the owners when they were out of touch with vessel owners or when funds could not be obtained from such sources, and it was necessary to make repairs to facilitate the safe operation of a vessel on a return voyage to its home port or to its country of registry.

On January 28, Cooper had the *Struan* moved to the Mission Rock area for unloading ballast, and on February 9, using the funds made available from his negotiation of the bottomry bond, had his vessel towed to Hunters' Point[8] drydock to be overhauled. Two weeks later, after substantial repairs, the *Struan* was refloated, and armed with a new survey by ship-rating bureaus, was towed up to Wheatport for loading.

Cooper had conquered his major problem—getting the ship put into first class sailing order. He now had a seaworthy vessel and a cargo which would provide an adequate if not generous fee for transporting it. But there was another difficulty. Under little known and sometimes misunderstood regulations, funds obtained through the use of a bottomry bond could be used only for actual repairs to the vessel itself. They could not be expended for ship's stores, rations for

---

[8] Named for the two Hunter brothers, Robert and Phillip, who were hired by the South San Francisco Homestead Company in 1849 as real estate agents to sell parts of a large tract of development land. They moved onto the property and built a house at a shoreside location that by 1853 came to be known as Hunters' Point. In 1865, after one of the new major owners had donated ten acres for the purpose, civil engineer A. W. Von Schmitt began construction of the drydock, using a crew of 15 men. The first ship to use the new dock was the sidewheel steamer *Ajax*, docked October 22, 1868. (From the San Francisco Naval Shipyard publication *Drydocker*, December 15, 1961).

the crew, or crewmen's wages. Cooper finally recognized that he would have to use his own funds to pay the costs of the trip home.

One of the things that attracted the *Struan* skipper's attention during the last few days before departure was the preparation being made to erect a staff on the top of the Merchants Exchange building from which a time ball would descend each day at exactly noon to enable shipmasters to set their chronometers accurately. A telegraph line from the Mare Island Naval Observatory would provide the necessary information when arrangements had been completed.

While he waited for the transfer from his own bank, Captain Cooper spent many daytime hours ashore, probably dining more than once on the tasty bay shrimp available almost everywhere in the area. These gourmet delights were caught by an estimated 1,500 Chinese fisherfolk in camps situated around the bay at locations where the shrimp were plentiful.

When he made some preliminary exploration of prices at a number of ship chandlers and provisioners (he would, after all, be spending his own money) he discovered that some were better equipped than others to furnish certain types of items at a good price. He would realize substantial savings by utilizing different suppliers. Perhaps his most tantalizing discovery was Ghirardelli's famous chocolate, in 12 or 24 pound bars, available in wholesale quantities at 21¢ per pound.[9]

Finally, on March 31, 1885, Starr & Company cleared the *Struan* through customs with 52,261 centals of wheat worth $67,939, destined for Great Grimsby, England. Cooper hired ten able-bodied seamen under somewhat unusual terms. Most of them were entered on the articles to receive one shilling—less than 25¢ U.S.—pay per month during the first two months of service, and five pounds per month after that time, suggesting that the captain had been required to advance funds to pay off their existing debts—probably at sailors' boarding houses, before he could ship them aboard. Two months loss of pay at five pounds (per month) was at then-existing rates of exchange almost the equivalent of $50 U.S. Board and room at residences for sailors was easily available at $5 per week, but hired seamen may not have been gainfully employed for about ten weeks, because so many ships were laid up as a consequence of the low freight rates.

Captain Cooper picked up the ship's papers at the British consulate on April 1 and sailed from San Francisco Bay the following day.

---

[9] During the Spring of 1885, the San Francisco *Merchant* carried an advertisement offering the product at that price.

Very little information is available concerning the *Struan*'s voyage back to England. She was spoken by another vessel on July 27, but that information was not reported to and published by Lloyds until the week of August 21 when the hailing vessel itself reached its next port of call. The sighting took place in 23°N latitude and 39°W longitude, almost exactly in mid-North Atlantic, close to the Tropic of Cancer. The *Struan* would have spent almost four months sailing down the west coast of North, Central and South America, around Cape Horn, and up the east coast to a position about five-sixths of the way to its destination.

Cooper brought his ship into Grimsby Roads September 3 and went ashore almost at once to take the Humber River ferry up to Kingston-Upon-Hull (the city of Hull) where he employed the firm of Joseph W. Hearfield and Lambert to file suit for himself and 22 other crew members against the owners of the ship *Struan* and her freight in the amount of £1900 (though he himself was a 8/64ths owner of the vessel) "for wages due to him and for disbursements made by him in respect of the said vessel and the remaining plaintiffs as the officers and crew of the said vessel claim for wages due to them in respect of their services on board the said vessel."

By this maneuver, Captain Cooper, knowing or suspecting that the major *Struan* owner, James Underhill Robertson, who held a 32/64ths share in the vessel, might be unable or unwilling to meet the burden of paying off the bottomry bond Cooper had signed in San Francisco within the five-day limit after his arrival at Great Grimsby, hoped to be "first in line" with his claim in the court battle he anticipated.

Although his Hull solicitors engaged Pritchard & Sons, No. 9 Gracechurch Street, London, to present their case for Cooper before the Admiralty Court, and the suit was filed in timely fashion on September 10, Cooper actually received no real benefit from his hurried filing. On September 14, Solicitors Rollit and Sons initiated an action on behalf of Wilson Meyer & Company, with whom Cooper had signed the bottomry bond in San Francisco. With their complaint they presented an affidavit signed by co-partner John Wilson, with a copy of the bottomry bond annexed, and by that action for the first time publicly disclosed the amount of same to be for 13,000 dollars and 50 cents ($13,000.50) with maritime premium of 29 1/4 per cent, to be payable "within five days next after the arrival of the said ship at her moorings at Great Grimsby." Payment both on the loan and the premium thereon was to be made on the basis of $4.86 of gold coin of the United States for every British pound sterling, resulting in a total indebtedness under the bottomry document of £3457 6s. In the same filing, holders of the bottomry bond "prayed for a warrant to arrest the ship *Struan*."

The very next day, Tuesday, September 15, 1885, acting under cover of judicial approval, Charles Alexander Burgoyne, a clerk in the office of Rollit & Sons, solicitors for Wilson Meyer & Co., went aboard the ship *Struan*, then lying at the dock at Great Grimsby in the County of Lincoln, and served his employer's writ of summons against the vessel by nailing the original to the mainmast for a short time and by nailing in its place a true copy when he removed the original.

At a hearing on September 22, the vacation judge permitted the bottomry plaintiffs to consolidate the two actions by paying "such wages of the officers and crew (excluding the master) of the *Struan* as are properly due together with one guinea[10] for their costs" and he directed that said plaintiffs should have first charge on the vessel itself and on freight charges due for transportation of the inbound cargo of wheat from San Francisco.[11] That action brought into the case Marshall and Sons, merchants at Great Grimsby, consignees and owners of the cargo aboard the *Struan*, who were represented by the same firm of solicitors, Pritchard & Sons, that was already involved with Captain Cooper's original claim. On September 28 the plaintiffs paid to the officers and crew of the *Struan* their wages up to that date, amounting to £1010 6s. ld. on the terms specified in the order made by the Honorable Mr. Justice Mathew. In his ruling it had been ordered that plaintiffs should recover the amount of wages expended in addition to any other amounts due them upon conclusion of the case.

On October 14, another firm of solicitors, Stokes, Saunders and Stokes of London, acting as agents for Bateson, Bright and Warr of Liverpool, entered an appearance as intervenors for Eills and Company, which firm held a mortgage on 32/64ths of the *Struan*'s ownership shares. That filing caused the judge to call for a report by the Registrar of Shipping, one of whose duties was to keep accurate and up-to-date records of the actual ownership of each share of every British-registered vessel, including those registered at Colonial ports. The registrar's report verified facts set forth in the Eills affidavit, specifically that James Underhill Robertson, who held 32 shares of the

---

10 Formerly a British gold coin worth 1 shilling more than a pound sterling, or 21 shillings. At the 1885 rate of exchange, equivalent to $5.10 U.S.

11 Approximately $23,000 U.S. or 4,750 British pounds, for the transportation of some 2,350 tons.

Ship *Struan* dating from January 17, 1881 at 11 a.m., had on February 17, 1885 at 3:25 p.m. mortgaged his entire one-half ownership in the vessel to William Eills and Barton William Eills, merchants of Liverpool, County of Lancaster, joint mortgagees, for the sum of £4000, on a mortgage dated October 3, 1883. The timing of that transaction is of particular interest when we recall that on that date the *Struan* was in Hunter's Point drydock at San Francisco, midway through major repairs, and Robertson had been the recipient of several telegraphic messages from Captain James Cooper pleading for funds to initiate required work to make the vessel seaworthy for a safe trip back to England. No record has been found to indicate that any part of the funds James Robertson obtained from Eills was ever sent to Cooper or otherwise utilized for the benefit of the vessel itself.

Another question arising from these facts is how James Robertson was able to obtain £4000 on a mortgage for half of an eight-year old wooden vessel that was only partly repaired (indicating a total value of some £8000) when the court-ordered appraisal and sale of the entire ship some six months later in a fully-restored condition, brought only about £4100, requiring that the court "bring the freight charges into court to satisfy other claims in the case."

Had it not been for the fact that Robertson had mortgaged his entire share of the vessel, the freights due for transportation of the grain cargo to Grimsby would just about have covered the amount due on the bottomry bond and the crew's wages for the trip home. Captain Cooper's recovery of the amounts he personally expended for ship's gear, supplies and rations, as well as his own salary, was prejudiced by his position as a part owner of the ship.

The last entry pertaining to the Louis A. Robertson ownership of a share of the *Struan* was his belated transfer, by a bill of sale dated September 7, 1885 at 11 a.m. (two days *after* arrival of the *Struan* at Grimsby) to someone having an address at 12 Utter Road, Prince's Park, Liverpool. Spilled ink on the original record makes identification of the intended transferee impossible, but enough is evident from under the inkspots to lead to the conclusion that a woman received the shares, possibly the widow of one of the previous owners. At the date of that transfer, it can reasonably be assumed that Louis was aware of the ship's presence at Grimsby, of the mortgage on the vessel that had to be paid within days, of the fact that major owner James Underhill Robertson had mortgaged his entire holdings for a £4,000 loan on February 17, 1885, and that there would undoubtedly be a major legal hassle over disposition of the *Struan* and its cargo. Transfer of his four shares may well have been his method of disposing of a minor share which he considered (correctly) might involve him in protracted and expensive litigation without much hope of realizing any return.

The *Struan* was condemned by the court, sold at auction by the Admiralty Marshal on November 27, 1885, and came into the possession of Andrew Gibson, a Liverpool shipowner who was no stranger to the ship, since he had, on at least one previous voyage, acted as agent for the vessel at various ports of call.

Chapter XI

Latitude, Longitude and Dead Reckoning

Using his sextant (introduced in 1731) and chronometer (invented in 1738) a competent navigator or shipmaster, in favorable weather permitting his observation of the sun and/or other celestial bodies, could in the *Struan*'s time have determined the position of his vessel within two miles of actual latitude and three miles of longitude. In accomplishing that exercise he would have utilized his considerable knowledge of mathematics (including trigonometry and the function of logarithms) and demonstrated his familiarity with the tables and other information contained in the *New American Practical Navigator*, compiled and published by Nathaniel Bowditch, an American astronomer, mathematician and navigator, in 1802 when he was only 29 years of age. For 150 years after its introduction, Bowditch's treatise was the bible for navigators all over the world. The U.S. Navy Hydrographic Office printed probably fifty editions during that time.

While a plotted position only two or three miles in error satisfied the requirements of navigation on the open sea in that era, a mistake of such magnitude could spell a vessel's destruction in a narrow waterway, near underwater obstructions, or along a coastline with strong currents or onshore winds. An erratic, or inaccurate chronometer compounded the problem of determining actual location, and in many cases led to destruction of the vessel, especially if coastal landmarks were obscured by fog or heavy snow. Because the chronometer's accuracy was so important to the safe operation of the vessel, many captains required one of their mates to wind the instrument regularly at the required interval and to report that action on a prearranged schedule at a certain time each day. If he received no such report by the appointed hour, the shipmaster could wind the chronometer himself and "have a chat" with the offending officer.

With the advent of radio communication in the first quarter of the 20th century, the problem of inaccurate chronometers was eliminated when the U.S. Naval Observatory began broadcasting accurate time signals at frequent intervals each day.

Within ten years after World War II, improvements in navigational devices enabled ships and planes to carry gyroscopic equipment known as "dead reckoning tracers" (DRT) that printed a continuous line on a chart to show the actual course being followed and distance traveled, as well as indicating the approximate latitude and longitude of the most recent position.

Construction of Loran radio navigation stations soon thereafter enabled pilots of ships or planes arriving over water[1] to plot their positions even in darkness, fog, or blinding snowstorm, by finding on the charts the intersection of signals from two radio stations, all with no dependence on the sun or stars.

Nowadays, anyone with the ability to change batteries or slide in a recharged battery pack can use a small hand-held instrument no larger than a cellular telephone to ascertain accurate latitude and longitude to within *50 feet* of exact location almost anywhere on the earth's surface.[2] Utilizing the Global Positioning System (GPS) originally developed for the U.S. Armed Forces, these small receivers incorporate satellite technology first tested in combat during the Gulf War.

Tracing the voyages of the *Struan* and other vessels mentioned in this book will be made easier for the reader if he or she has a basic understanding of the manner in which latitude and longitude are measured and utilized to record a specific position anywhere on the land or sea surfaces of the planet.

Latitude

In the measurement of latitude, an imaginary circle is drawn in an east-west direction around the center of the earth. This line is called the equator, and measurements north and south from the equator are made in degrees, minutes and seconds, though for most purposes the seconds are usually omitted. For example, 31° 22' N is sometimes

---

[1] Loran-based signals were limited in range when transmitted over land areas. Signal strength was also affected by daylight. Maximum effective distance for receipt of Loran signals occurred in hours of darkness.

[2] Some of these tiny devices are also equipped to calculate and display height above sea level in addition to latitude and longitude.

written 31.22 N because at one time most type fonts used by hand-set newspapers and periodicals did not provide a degree symbol in their alphabets. One minute of latitude is theoretically equal to one nautical mile or 6,080 feet (in Great Britain), formerly in the U.S. 6,080.27 feet, and by generally recognized international usage 6,076.115 feet, which has now been officially adopted in the U.S. Since there are 60 minutes in each degree, 60 nautical miles represents close to one degree of latitude, or about 69 land (or statute) miles of 5,280 feet, measured along the great circle route, i.e., along the imaginary line passing through the north and south poles and the point whose location is being determined.

If the earth were a true sphere, as mariners not too long ago generally accepted it to be, the "one minute of latitude equals one nautical mile" rule of thumb would prevail without discussion. In fact, a respected British encyclopedia in 1771, in an article explaining latitude and longitude, made no reference to the slight variation in the length of a degree of latitude[3] resulting from the earth's actual shape.

Later recognition of the earth as an oblate spheroid (ellipsoid), in which the planet is flattened at the north and south poles to the extent that the radius at those points is some one hundred miles less than that at the equator, caused navigators as well as world geographers to realize that slight discrepancies in the actual length of a degree of latitude do exist and should be considered. Lengths of a degree of both latitude and longitude in all 90 latitudes, correct to three decimal points, as measured in nautical miles, statute (land) miles, and meters were enumerated in an 1866 publication.[4]

Its table No. 4 discloses that a degree of latitude measured in the area of 45° N or S latitude has been computed at 59.965 nautical miles while the length of a degree in 32° N or S latitude is 59.831 nautical miles—not a great difference at a little over 1/8th of a mile or 815 feet. Minor variations that do exist in the length of a degree of latitude are most pronounced in a comparison of measurements made at the equator (59.661), and at the poles (60.267), in nautical miles.

Parallels of latitude are all exactly that—imaginary lines drawn

---

[3] A Society of Gentlemen in Scotland, *Encyclopedia Britannica; A Dictionary of Arts and Sciences* (Edinburgh: Printed for A. Bell and C. Macfarquhar at the printing office of Colin Macfarquhar, Nicolson Street, 1771) III p. 372.

[4] From the Clarke Spheroid, published usually in alternate years at the U.S. Navy Hydrographic Office as part of the American Practical Navigator by Nathaniel Bowditch.

parallel to the equator so that every point on each parallel is the same distance from the equator as every other point. Measurement of latitude (either north or south from the equator) can be from 0° to 90°, 0° being the equator and 90° either the north or south pole. No measurement of latitude exceeds 90°, since a measurement from the equator to the same pole from a point halfway around the earth (180° of longitude away from the given point) uses the other 90°, consuming 180° and effectively providing a north or south measurement structure for one-half of the planet's surface, or one hemisphere. The other hemisphere utilizes another 180° in the same manner, thus involving a great circle's entire 360° in measurement of latitude.

The 45th parallel of north latitude (halfway from the equator to the north pole) for instance, passes through the northern outskirts of Salem, Oregon, U.S., the state capital. On the opposite side of the planet the same 45th parallel bisects Lake Aral in what was formerly the U.S.S.R. Similarly, the 32nd parallel of south latitude runs through Rattnest Island (off the Indian Ocean coast of Western Australia) between Perth and Fremantle, and on the opposite side of the world passes just south of the city of Los Vilos on the South Pacific coast of Chile.

An easily arranged visual illustration of the equator and parallels of latitude may be of help. If a tennis ball were placed in an empty drinking glass the rim of which grasped the ball at its exact midpoint, the line delineated by the rim of the glass encircling the ball would represent the equator. If a series of smaller inverted glasses were carefully held against the exposed half of the ball in the same axis as the original glass, lines defined by the edges of each succeeding smaller glass and drawn around the protruding half of the ball would represent parallels of north latitude. As the inverted glasses decreased in size, each line drawn as a parallel of latitude would become shorter than the previous one.

Longitude

Measurement of longitude involves a different system. The Greenwich meridian, which runs through London, England, is the baseline. On the exact opposite side of the world, in the Eastern Pacific Ocean, is the 180th meridian, also generally recognized as the international date line, which, although it coincides with the 180° meridian in many areas, wanders back and forth in several locations, most often to avoid instances where two different time zones might be confusing in a relatively small geographic setting, as around groups of islands.

Each of these meridians, and the 358 others in between them, are *great circle routes* passing through both poles. Longitude is measured east or west of the Greenwich meridian, up to 180° in each direction, using a total of 360°, which have been divided into 24 time zones, each containing some 15°.

Since longitude is delineated by north-south great circle routes drawn through the locations to be measured, it is evident that moving 45° of longitude (from east to west, or west to east) at the equator requires a much longer trip than if the journey were made close to one of the poles. Thus, while a degree of *latitude* is always approximately 60 nautical miles in length, a degree of *longitude* varies with the latitude in which it is measured. From a table[5] published in 1771, for instance, a degree of longitude at the equator is 60.07 nautical miles, at 45° N or S latitude is 42.43 nautical miles, and at 80° N or S the distance is reduced to only 10.42 nautical miles. It would then be theoretically possible, near one of the poles, to move through 360° of longitude in just a few hours.

In the maritime community, when latitude and longitude were being specified, it was customary to give the latitude first, then the longitude. In news articles and periodicals one hundred years ago, "lat. 45° 11' N, long. 134° 12' W," or simply "45.11 N, 134.12 W," would have assumed that most readers understood that N or S indicated the latitude, and E or W the longitude.

THE LOG

One kind of ship's log, as differentiated from the official written record of the vessel's voyages, port calls, etc., is a combination of mechanical devices assembled for measuring the speed of a ship. Although early versions of patent logs, designed to rotate in the water astern of the vessel and record speed on a dial, were available to mariners as early as 1835, they were not well received by shipmasters of the time. Most deep-water sailing ship captains preferred the old-style logs with which they had been familiar since their days as apprentices. It was not until late in the 19th century that improved models of patent logs were able to overcome many

---

[5] A Society of Gentlemen in Scotland, III p. 372.

mariners' skepticism or outright suspicion about these new methods for measuring speed.

The simple (often shipboard-made) log of the *Struan*'s time consisted essentially of three operating parts, including (1) the board, sometimes called a chip; (2) the log line; and (3) the log reel. It was also necessary to have available either a 28-second sand glass or access to a pocket watch or chronometer with a second hand.

The board was usually triangular in shape, weighted to float in a perpendicular position, and attached by a bridle to the log line in such a fashion as to cause the board to exert maximum resistance to the water as the ship pulled away from it. (The object was to have the board stay in the same position where it landed upon being thrown overboard, and to resist any tendency for it to be towed by the ship). The log line itself was usually 600 or 700 feet long (or longer if the vessel's master expected to exceed 15 knots—15 nautical miles per hour) with a knot tied in it every 47'3" and was carefully coiled on the log reel to avoid tangling as it unwound over the stern.

To measure a vessel's speed, the board (or chip) was thrown overboard, the exact time noted, and the log line kept slack by rotating the reel as the line ran out over the stern. In good light, the knots in the log line could be counted as they passed overboard. In some instances each knot in the log line had been daubed with a bright colored paint to make it more visible. At night or under severe conditions, a seaman wearing a glove could count the knots as they passed through his hand. The number of knots that were counted in a 28-second time period after the board hit the water was the speed of the vessel in nautical miles. If a very accurate reading was required, the person holding the timing device called out "hold" at the very expiration of the 28-second interval, and the sailor handling the log line grasped it firmly to prevent paying out any more line. Then the distance in feet from the last knot observed going overboard to the point where the line was "held" could be compared to the 47'3" distance between knots to make available a fraction of a knot. A distance near 24 feet, for instance, indicated that a half knot should be added to the vessel's speed as determined by full knots. Upon completion of the speed test, the log line could be pulled aboard and carefully rewound on the log reel ready for its next use.

This ingenious but simple device depended for its accurate operation on the relationship of the distance between knots on the log line to the length of a nautical mile (6,080 feet), and a similar relationship of the 28-second time frame to the number of seconds in a full hour:

47.25 feet	=	28 seconds
6,080 feet (prox.)		3,600 seconds

Dead Reckoning

When weather conditions prevented a mariner in the 1880s from "shooting the sun" or "taking a fix" with his sextant, he could navigate for a limited period of time through use of dead reckoning—the determination of his vessel's position from a record of the courses sailed and the distances traveled applied to a known starting point—all adjusted for currents, wind and drift. The courses sailed were those specified by the navigator for the helmsman. Distances traveled were computed from a recap of the hours under way at the speeds indicated by the log as previously described. The known starting point was usually the last position established by actual celestial observation. At the beginning of a voyage, the port of departure's known latitude and longitude were often used.

Positions at sea established by dead reckoning were always subject to mistake, and the likelihood of error increased substantially as the length of the dead reckoning cycle was extended. Minor deviations from real location after sailing two or three days by dead reckoning became major variations after a week or more of not being able to make an actual observation.[6] Courses actually sailed could vary slightly from those intended for a variety of causes, including compass error or a steerman's inattention. Speed measured at one time of day could vary from hour to hour even on the same course with the same apparent wind, putting in error the distance traveled computations. Too generous or too spartan an allowance for drift, wind slippage, or unrecognized currents all contributed to the possibility of error. Few mariners, after finally being able to make an actual observation following days of sailing by dead reckoning, found their vessels to be in the position indicated by the dead reckoning calculations.

On some ships, two different officers made dead reckoning computations, not only to average the errors, but to determine which individual was consistently most accurate in determining positions. A reputation as a capable navigator particularly accurate at dead reckoning computations never hurt any young officer's chances for a future

---

[6] See middle of page 140.

command of his own. Because no shipmaster ever knew with certainty in advance when he might be forced to navigate by dead reckoning, entries of courses sailed and estimated distance on each course were carefully recorded for possible future use as required.

Distances traveled as recorded daily in a ship's log book, as well as such measurements from port to port, were always expressed in "over the bottom" nautical miles. Although a sailing vessel's taffrail log might have indicated a speed through the water at ten knots for an entire 24-hour day, (240 knots in 24 hours) the zig-zag courses traversed to take advantage of prevailing winds and currents might have reduced the actual progress along the intended course to only about 120 nautical miles. This mathematically computed measurement would be recorded in the ship's official log as distance traveled in the 24-hour period. For a square-rigged ship like the *Struan*, the over the bottom distance averaged only about 50 percent of the actual miles sailed on an ordinary voyage. A skipper could expect to improve his long-term average by more than a few points only if the wind blew from dead astern or from the port or starboard quarter for an extended period while the vessel headed directly along its intended course.

Chapter XII

Two Cargoes of Cotton, Two Captains of Commerce

When Andrew Gibson, a shipowner in Liverpool, County of Lancaster, England, came into possession of the *Struan* in 1885 by purchase at an auction sale conducted by the Lord High Marshal of the Admiralty Court in London, he appointed Samuel Lemon as the new master. Lemon had been a mate in the *Struan*'s crew on every voyage since its launching except the initial passage from Saint John to Liverpool. He first signed on as the vessel's second mate for the trip originating at Liverpool September 13, 1877, and by 1880 was sailing as first mate. With more than seven years experience in the ship, he was the logical choice. Lemon had been born in Bangor, County Down, Northern Ireland in 1851 and obtained his certificate of competency as a shipmaster (master mariner) at Dublin in 1882.[1]

Whether James Cooper was considered for command under the new owner is questionable. Cooper had captained the *Struan* for eight full years, had just returned from a difficult three-year voyage during which his vessel experienced serious storm damage and was becalmed for an extended period, and he had been unable to obtain funds from the owners to make necessary repairs to the ship. At 55 years of age, he may have decided to retire from active duty at sea.

He may not have been offered the post, however, because of his having brought a suit against the ship on behalf of himself and the officers and crew, for unpaid wages, though no criticism of his action

[1] Samuel Lemon was mate on the *Struan* while it was in port at Dublin for almost a month in April-May 1882. It can reasonably be assumed that Captain James Cooper permitted his first officer time away from his regular duties to take the master's examination.

in that matter could have been considered reasonable under the circumstances. (See top of half of page 88.)

Gibson arranged to have a new certificate of Registry issued as of July 2, 1886, enumerating the *Struan* as the 57th vessel registered at Liverpool in 1886, with himself as the sole owner of all 64 shares. In the meantime, Captain Lemon, apparently operating on the original Saint John vessel registration, signed a new crew aboard on December 14, 1885 and set sail for New Orleans on the 17th from Grimsby in ballast. The new crew included only one member of the previous *Struan* ship's company, boatswain's (bos'n's) mate John Bragg.

Progress of the voyage can be followed by references to the *Struan* in New Orleans newspapers of the time.

From the New Orleans *Daily Picayune* (since the early 1990s the *Times-Picayune*) Sunday, February 14, 1886:

> The British ship *Struan*, 1473 tons net, Captain S. Lemon from Grimsby, England, Dec. 17,[2] in ballast, at present waiting orders at Port Eads was entered at the Custom House yesterday by Messrs. Silas Weeks & Co. Captain Lemon arrived in the city yesterday morning. He reports that his vessel on sailing had thick and foggy weather down to Beachy Head,[3] followed by fine weather to Jamaica, thence calms and easterly winds to Cuba, where strong variable winds and heavy rains were encountered, followed by SW winds to the Passes.[4] Northwest of Cuba she passed a large quantity of deals.[5]

Port Eads was a project headquarters, work camp and pilot station on the South Pass channel at the mouth of the Mississippi River. It was established in 1875 by and named for James Buchanan Eads (1820-1887), the self-taught engineer who planned, designed and constructed the jetties that eliminated shoaling in the channel, thereby bringing to New Orleans world recognition as a major deep-water port.

Eads had made a fortune by the time he was 37 years of age

---

[2] This date indicated the date of departure on the current voyage.

[3] A promontory (cape) on the south coast of Sussex, England three miles west of Eastbourne. The chalk cliffs rise to a height of 564 feet. A lighthouse visible from the sea for 23 miles was erected on the west cliff in 1823, and the Head soon became a major landmark for mariners sailing the Dover Strait and the English Channel.

[4] The three channels through the Mississippi River delta were the North Pass, the South Pass and the Southwest Pass, collectively known as "the Passes".

[5] Such flotsam would have indicated either a vessel's loss of its deck cargo, or the breaking up of a vessel and release of the deals (lumber) in the hold.

salvaging, repairing and rebuilding sunken steamboats throughout the entire Mississippi River system, often utilizing the diving bell mechanism he had developed. During the Civil War he turned out river ironclads for the Northern forces.

His first claim to fame was the construction of a steel arch bridge with a 520-foot span across the Mississippi at St. Louis, after 28 leading civil engineers had stated that such a feat would be impossible. During construction, exceptionally hot weather expanded the steel members to be connected at the center of the span, but Eads had his men pack the sections with ice-filled grain bags and continued the project on schedule.

After the Civil War New Orleans merchants and shippers complained that the shallow channel at the mouth of the Mississippi was impeding their access to world commerce. The Army Corps of Engineers proposed an expensive ship canal that might have entailed closing the river to vessels for most of the construction period.

Eads countered with a plan to build jetties at the mouth of the river in one of the three passes to confine the current to a narrower but deeper channel, forcing the thousands of tons of mud that came down the river each year out into the Gulf of Mexico instead of depositing most of it in the channel. He took a Congressional delegation to the site, explained his plans, and finally went before Congress with a "no cure, no pay" offer to produce and maintain a 28-foot channel at his own expense if the government would pay him $10,000,000 if he succeeded; it would owe him nothing if he failed. After some negotiations, he was awarded a contract on his own terms with a smaller sum of money because the jetties would be built on the narrower South Pass, than on the wider Southwest Pass where he had first intended to build. On completion of the work in 1879, the depth of the channel at average flood tide measured 30 feet; Eads had conquered the Mississippi and received his money.[6]

In port at New Orleans when the *Struan* arrived were 28 steamships, 16 sailing ships, 24 barks, one brig and 7 schooners. The main topic of conversation along the waterfront and in business houses uptown was the "ruinously low price of cotton" (only 6 1/2 to 8¢ per pound). Newspapers were advertising the Mardigras presentation of the *Krewe of Proteus* for the night of March 9, and Kersand's

---

[6] Ray Samuel, Leonard V. Huber, and Warren C. Ogden, *Tales of the Mississippi* (Pelican Publishing Co., Gretna, 1992) pp. 57-63. Eads died of pneumonia at Nassau, New Providence Island in the Bahamas on March 8, 1887.

Minstrels offered evening performances at 10¢, 20¢, or 30¢ admission. Mule races were scheduled for April 9.

From the New Orleans *Daily Picayune*, Tuesday, February 18, 1886 (page 8):

> The British ship *Struan*, 1473 tons net, Capt. S. Lemon, from Great Grimsby, England, Dec. 17, in ballast, which has been waiting arrangements for disposal of ballast[7] at Port Eads, arrived up in tow yesterday morning and was anchored at the Point. She will load for Liverpool. Silas Weeks & Co. are her agents.[8]

From February 19 through about April 2, the *Struan* was moored at the foot of First Street in New Orleans, and was then moved to the Second Street dock for loading.

From the New Orleans *Daily States*, April 15, 1886:

> To clear today for Liverpool, the *Struan*, Lemon, Silas Weeks & Co., carrying 5,370 bales of cotton[9] and 3,080 pcs of staves.[10]

---

[7] Vessels planning to load a lightweight cargo like cotton required additional ballast deep in the hold to give the ship stability in strong winds and heavy seas. A captain arriving with his ship in ballast would be hesitant to unload it all until he knew for certain what cargo he would be loading.

[8] Silas Weeks and Company, with offices at 48 Carondelet Street in the French Quarter, was one of a number of established shipping agents in the city, and had been one of the 99 firms included in the first telephone directory ever issued in New Orleans—a one-page listing made public by the New Orleans Telephone Exchange in December 1879 after introduction of the service upon receipt of a license from the American Bell Telephone Company. Business rates were $2 per month, and no phone numbers were assigned.

Subscribers were instructed when initiating a call to "give a long ring on the bell crank," and when the operator signaled back to "remove the telephone (receiver) from the hook and give the name of the subscriber wanted." When called by the exchange (for an incoming call) the customer was advised to "give a ring-back, remove the telephone from the hook, and say Hello! Hello!"
Leonard W. Huber, *New Orleans—A Pictorial History From the Earliest Times to the Present Day* (Pelican Publishing Co., Gretna, 1991) p. 319.

[9] Bales of cotton varied greatly in size, depending to some extent on the power and design of the press utilized. The range of common dimensions included a normal length of 54 to 58 inches, a width of 42 to 46 inches and a thickness of 24 to 30 inches. Bale weights usually averaged in the neighborhood of 500 pounds.

[10] Shaped wood members intended to be assembled into barrels, kegs or tubs, probably made of Louisiana oak.

It is believed that staves were carried for their utilitarian value as dunnage, cribbing, bracing, etc. to hold the cotton bales firmly in place during the voyage, not as a revenue producer. At the then-existing freight rate (£10 per M pcs.) the 3,080 pcs of staves would have netted the ship only about $150 in transportation fees.

In spite of having to replace nine members of the crew who deserted in New Orleans, Lemon brought the *Struan* into Liverpool on May 28, 1886, having made the eastbound crossing fully loaded in about six weeks, whereas the westbound trip had required almost two full months in ballast. Because of more favorable winds and currents, the eastbound voyage from North American ports to British destinations normally required less time than did a trip in the opposite direction. The voyage back to England in this case could be considered about four hundred miles shorter, too, since the *Struan* did not have to sail through the Dover Strait and part way up the east coast of England, but went instead to Liverpool, a major port on the country's west coast.

Taking only a few days more than a month for unloading the cotton from New Orleans, replenishing supplies and taking aboard the outbound cargo, the *Struan*, still under command of Samuel Lemon, with a new crew, left Liverpool July 5, 1886 for Rio de Janeiro, Brazil with 2,436 tons of coal. She was observed off Lynas Point, Wales on July 6th and reported near Tuskar Rock, Ireland the next day, but was apparently not reported or spoken again during the entire trip to South America.

The *Struan* arrived at Rio on September 7 and was directed to the Gambôa area for unloading.

Publication of vessel arrivals the next day included the entry "from Liverpool, 62 days, British full-rigged ship *Struan*, 1,473 tons, Captain Samuel Lemon, crew 23, cargo—coal to Rio de Janeiro Gas Company."

On September 11th, customs officials inspected the ship and its cargo, and on the same date the vessel was included in a published list of "foreign vessels anchored in this port." That information, together with the fact that "unloading in bulk," begun on September 14, took until October 20 to complete,[11] leads to the conclusion that the vessel probably discharged its cargo into lighters at its anchorage. Such a method of unloading would also help to explain why, in spite of Rio's reputation at the time as an ideal place for sailors to "jump ship," no crew members left the vessel. It can be reasonably assumed that the location of the *Struan*'s anchorage was far enough from shore to discourage even any swimmers bent on desertion.

The *Struan* was cleared for departure on October 23, and after

[11] "*Jornal do Commercio*" (RJ) -Microfilms: CPR-SPR 1 (153) and (154) at the Biblioteca Nacional, Avenida Rio Branco 219, Rio de Janeiro, Brazil. (Translated from the original Portuguese).

leaving able seaman Patrick Murphy from Cork, Ireland sick at the Rio hospital, took on stone ballast before leaving on October 27 with the announced destination of "S.W. Pass" (one of the channels through the Mississippi River delta).[12]

From the New Orleans *Daily Picayune*, Sunday, December 12, 1886 (some Sunday editions contained as many as 16 pages):

> The British ship *Struan* 1474 tons, Captain Samuel Lemon, from Rio de Janeiro in ballast, consigned to Silas Weeks and Co., now at Port Eads, was entered at the custom house yesterday.
>
> She reports having sailed on the 27th of October and had strong SE winds to about latitude 10°S, thence to the equator had NW winds through the islands and until off the Passes, where she encountered a strong NW gale with rough seas, whirlwinds and many waterspouts in the vicinity, 18 of them being visible at one time. On November 19, in latitude 8°S, longitude 25°W spoke the British ship *Lord Torridon* of Glasgow, bound south. She wished to be reported. The *Struan* is due in port on Tuesday and on her arrival will load for Havre.

From the New Orleans *Daily Picayune*, Thursday, December 16 (page 6):

> The British ship *Struan*, 1474 tons, Captain Samuel Lemon, 41 days from Rio in ballast, previously reported below waiting and its report published, arrived up in tow yesterday. The *Struan* was landed at position 32, first district, where she will load for Havre.

Mariners arriving at New Orleans about the time the *Struan* docked on this trip found the city in turmoil over the newly passed Sunday closing law. Tavern keepers were being arrested and penalized with substantial fines for opening their premises or even dispensing their products through back doors on Sundays as an accommodation to regular customers. Meanwhile, arguments raged within the shipping fraternity either in favor of or against a suggestion to mount a steam syren [*sic*] on the jetty at Port Eads.

Some ocean freight rates from New Orleans in effect at the time included:

---

[12] This destination, declared at the time of departure, was garbled in telegraphic transmission to London, and was printed by *Lloyds Weekly Index* as "South Pacific", which would have required the vessel to sail south from Rio instead of north.

Oil cake[13] to the United Kingdom, 22 shillings per ton;
Staves to Liverpool, £10 per thousand pieces;
Grain to Liverpool, 6d (6 pence) per bushel;
Sugar to New York, $3.50 per hogshead (a large cask holding between 63 and 140 gallons);
Cotton to Liverpool, steam 12-32d, sail 21-64d per cwt.[14];
Cotton to New York, $.25 per cwt.;
Rice to New York, $.50 bbl.;

Key: British Coinage in 1886-87:

12 pence	=	1 shilling
20 shillings	=	1 pound
1 pound (symbol £)	=	US $4.94
1 penny (symbol d)	=	US $.02 (approximate)
1 shilling (symbol s)	=	US $.25 (approximate)

The British ship *Struan*, Captain Lemon, officially cleared New Orleans January 20, 1887 carrying 5,024 bales of cotton and 2,040 pieces of staves, bound for Havre, France; agents Silas Weeks & Co.[15]

The first published indication that the *Struan*'s voyage to Havre from New Orleans was other than a routine one came in a front-page report dated February 17, 1887 from Jacksonville, Florida:

> The British ship *Struan* from New Orleans for Havre with cotton for that place, is ashore in the Gulf of Mexico, sixty miles west of Cedar Key. The cargo of 5024 bales of cotton and the vessel itself are insured.[16]

It was later learned that Captain Lemon had gone mad and through incompetence in that condition put the Struan aground before he was finally replaced by the second mate, Samuel Pollock. On

[13] A mass of seeds and/or pulp remaining after the extraction of oil from cotton-seed, used by some farmers as livestock feed, by others as fertilizer.

[14] The disparity in freight rates charged by steam vessels as compared to sailing vessels probably resulted primarily from the fact that there were not enough steamships available to transport all the cotton that growers needed to ship. Cotton farmers who could not contract with a steam-powered vessel to take their crop to market were willing to pay the higher tariff asked by sailing vessels in spite of the delay occasioned by longer transit times.

[15] The New Orleans *Daily Picayune*, January 21, 1887.

[16] The New Orleans *Daily Picayune*, February 18, 1887 (front page).

the same day, arrival of another vessel, which was to have a part in the *Struan* story, was noted in the daily press.

From the New Orleans *Daily Picayune*, February 17, 1887 (page 6):

> The British ship *Mohur*,[17] 1413 tons, Captain John Scott, from Liverpool, January 8, with a cargo of salt consigned to Marnon Bros., vessel to Silas Weeks and Co. has arrived at the head of Sixth Street. She reports having strong winds and gales during the first part of the trip, followed by pleasant weather to the Passes.

Representatives of the ship's agents at New Orleans, Silas Weeks & Co., somehow managed to send word to the *Struan* to await the arrival of the steam tugs *Echo* from Pensacola and *Elmer E. Wood* from New Orleans, which had been dispatched by C.L. Labaron, agent of the British Lloyd and New York underwriters, on the basis of regular towage contract arrangements.

The mystery of where the *Struan* had been for the three-week period following its passing out into the Gulf of Mexico from Port Eads about January 23, 1887 until February 17 when reports began to be received concerning her grounding near Cedar Key cannot be solved using existing information.

When he learned at Pensacola, Florida of the *Struan* grounding near Cedar Key, the captain of the steamer *General Stafford* took his ship to the scene on February 18 to offer assistance, which was declined.[18] Upon their return to Pensacola, the *Stafford's* officers reported the *Struan* to be ashore about fifty miles west of Cedar Key, in 15 feet of water, on a soft bottom, about four miles offshore. The stricken vessel's condition was described as heading inshore, no anchors out and every stitch of canvas set. Four of her crew had deserted and no effort had been made to lighten the ship, which was "350 miles off course and drawing 17 feet."[19]

Samuel Pollock, then—who sailed as second mate from Liverpool in July—having witnessed over a three-week period Captain Lemon's apparent insanity and resultant inability to keep the *Struan* on a prop-

---

17 The *Mohur* was a 207-foot wooden sailing ship of 1,413 gross tons, built at Maitland, Nova Scotia in 1874; assigned official number 66524, signal letters WSRJ, registered in Liverpool, owned by A. Gibson.

18 Had the *Struan* accepted the offer of aid from the *General Stafford*, a salvage claim from that vessel might have resulted in a much higher cost to the insurance carriers than that of hiring the two tugs.

19 The New Orleans *Daily Picayune*, February 22, 1887 (page 6).

er course, must have realized by about February 14 that he should confine Captain Lemon and usurp control of the *Struan* on the grounds of Lemon's being unable to exercise good judgment and unfit to command by reason of insanity. But, having no other officer aboard to share responsibility or bear witness about the captain's condition, he hesitated and delayed taking action for too long. It is not clear from available records whether Pollock finally sent the captain ashore in a ship's boat or whether Lemon remained aboard until the arrival of the tug *Elmer E. Wood* from New Orleans, at which time he would probably have been landed by one of the tugs.[20]

Captain Saunders of the *Echo* reported on his return to Pensacola that his vessel reached the stranded *Struan* about 12 hours before the *Elmer E. Wood* from New Orleans arrived on the scene, and that he very likely could have pulled the *Struan* free of the sandbar at that time had Captain Lemon, whom he described as "*non compos mentis*" (not of sound mind), not insisted that his vessel be pulled farther on shore instead of toward the gulf, requiring the *Echo* to stand off until arrival of the New Orleans tug with a representative of the ship's agents aboard, who had authority to relieve the *Struan*'s master of command.[21]

Succession to the position of master of the *Struan* makes an interesting story. First mate William Bryant of Exmouth, who had signed aboard July 5, 1886 at Liverpool, was paid off at New Orleans on December 20, 1886 only a week after arrival there, and before loading of the cargo had begun. Available records indicate that he was discharged under a "by mutual consent" agreement between Bryant and the captain. Such an arrangement in many instances, especially between officers, indicated the existence of a personality clash or inability to "get along together." Second officer Samuel Pollock, from Bangor in County Down, Northern Ireland—the same city as Captain Lemon listed as his birthplace—was then apparently promoted to first officer, though no indication of his promotion or adjustment in his pay was made on the ship's articles.

A new second mate was employed three days later. That officer,

---

[20] Testimony at the May 1887 Belfast investigation regarding Captain Samuel Lemon's actions before and during the grounding(s) included evidence that the captain had been put ashore at a small settlement to prevent further damage to the vessel and cargo and danger to the crew.

[21] Captain Saunders' comments were recorded in the February 23, 1887 edition of the *Pensacola Daily Commercial* on the second day of its change to a daily publication from a three-times-a-week journal.

one I. Perring, along with four of the five able seamen employed at the same time, deserted soon after, though the exact date was not indicated in the crew list as it should have been. While desertion was common from vessel crews at the time, desertion by an officer was something of a rarity. Owners seeking mates and captains for their vessels were sure to weigh the dependability factor in an aspiring applicant, and a record of desertion was almost certain to be prejudicial to anyone whose record reflected such an action. With that in mind, it might be appropriate to wonder if the newly-hired second mate—perhaps even the seamen, too—could have detected, as complete strangers to Captain Lemon, some hint of an impending breakdown that was perhaps not immediately discernible to those who lived and worked with him on a daily basis.

None of the several endorsements to the *Struan*'s articles made by the British consulate at New Orleans in the period December 15, 1886 to January 21, 1887 were dated, which omission makes difficult the task of determining the sequence of events at that port. Since there exists no record of another second mate being employed while Lemon was still in command of the *Struan*, it can safely be assumed he sailed on or about January 21 without a second officer, but with 5,024 bales of cotton and 2,040 pcs. of staves as cargo.

On the previous trip, Lemon had the vessel halfway back to England within three weeks of leaving New Orleans. In this instance, he took three weeks moving less than 200 miles measured in a straight line. Some of the time may have been spent getting the vessel out of other minor grounding situations. In the extract of evidence developed at the investigative hearing convened to examine his actions in the matter, there is a reference to the *Struan*'s going aground *four times*; the fourth time apparently being the grounding sixty miles west of Cedar Key which required tugs to refloat the ship.

That extract, reprinted by the *Shipping Gazette Weekly Summary*, May 20, 1887, was copied almost verbatim from a news article in the *Belfast News Letter* of Wednesday, May 18, 1887:

> Local marine Board Inquiry: On May 17 an inquiry was held by the Belfast Local Marine Board into the conduct of the Master of the *Struan*, Samuel Lemon, while off the coast of Florida in February last. Mr F.J McCarthy, R.M. presided and Mr McTear appeared for the Board of Trade. From the evidence it appeared that the *Struan* left Liverpool about July 7 last, and proceeded to Rio Janeiro. They afterwards went to New Orleans where they arrived about December 16 or 17. A cargo was taken on board,

> and while off the coast of Florida the vessel went aground on four occasions. At this period the captain showed signs of madness, and all firearms etc had to be taken from him. Eventually, he was sent to a neighbouring town, the ship being taken care of by the mate. The captain was then conveyed back to Belfast, and placed in the District Lunatic Asylum. Dr. A.S. Merrick having been examined as to Lemon's condition at present, the Board cancelled his certificate until such time as he recovers....

During the night of February 21, the tugboats *Echo* of Pensacola and *Elmer E. Wood* of New Orleans, in a joint effort, were able to refloat the stranded *Struan*. Together they towed the ship to Pensacola where the *Echo* held the stricken vessel offshore in deep water while the *Elmer E. Wood* went in for coaling.[22] Subsequently, the *Elmer E. Wood* towed the *Struan* back to Port Eads, where an examination of the hull by both divers and marine surveyors was initiated. Having been inspected and declared seaworthy, the vessel was free to proceed on her trip under command of Captain John Scott [formerly] of the British ship *Mohur*, then in port at New Orleans.[23]

The *Struan*'s official crew list for the voyage begun at Liverpool July 5, 1886 contains the following notation *written in red crayon* opposite Captain Samuel Lemon's name:

> Left at New Orleans insane 26/2/87 in charge of ship's agents, Silas Wicks [*sic*] and Co. who would send him home to England. Official Log p. 23-(?).

The illegible initials inscribed at the end of the entry are believed by this author to be those of the *Struan*'s new captain, John Scott, although they could be those of any person having access to the ship's papers—one of the mates, for instance, or a consular officer.

Lemon was transported without delay from New Orleans back to Northern Ireland, where, on March 28, 1887 he was admitted to The Belfast Insane Asylum as an "urgent case" and his condition diagnosed as "Dementia Primary," supposedly caused by fright (no explanation offered). It was also noted that he suffered from Bronchitis upon admission. He survived only about six months after entering the institution, and died October 4, 1887.[24] His passing was noted in the

---

22 The New Orleans *Daily Picayune*, February 24, 1887.
23 The New Orleans *Daily Picayune*, February 27, 1887.

minutes of the October 10, 1887 meeting of the Board of Governors:

> Deaths recorded in the previous month, 3 (including)
> Samuel Lemon, 35, Nervous Degeneration, ill four months in hospital."[25]

A short note included in his file under "Observations" recorded the name James Lemon, 30 Powerscourt Street, Belfast, with no explanation. It could reasonably be assumed that the joiner (carpenter) living at that address until about 1890 was a brother, and had been listed as a family member or next-of-kin to be notified in an emergency, though Captain Lemon's wife is believed to have taken Bangor quarters in 1887 in a house with a garden at 4A Fishers' Hill (now Victoria Road), staying there until 1890.

At all times during his service as a mate aboard the *Struan* (from late 1877 through 1885) Samuel Lemon had availed himself of the British government's allotment plan for mariners, by which a portion of a seaman's earnings could be automatically paid directly to dependents by arrangement with the vessel owner. Usually half, or more than half, of Lemon's pay was remitted to family members under the allotment plan. Second mate's pay was six pounds (just under U.S. $30) per month, and when he sailed as first mate his pay increased to eight pounds (just under U.S. $40). A captain's pay scale did not appear on the ship's articles, and was privately arranged between the vessel owner and the master mariner involved.

Early in his career, allotments were made presumably to his mother, believed to have been Mrs. Mary Lemon, who operated a grocer's and haberdasher's shop through the 1860s, 70s and into the 80s at 59 Main Street in Bangor.[26] Later, after his marriage to Mabel Campbell, daughter of Hugh Campbell, the allotted portion of his earnings went to her. Hugh Campbell died September 20th, 1889, aged 82, and that information is inscribed on his tombstone in the Bangor graveyard. When Samuel Lemon's wife passed away, her father's tombstone was relettered, to read:

> also Mabel Lemon, who died 29th July 1915, aged 62 years,

---

[24] Public Record Office of Northern Ireland, Admissions Book, 1887, Belfast Insane Asylum, HOS 28/1/3/2.

[25] PRONI, Proceedings Book, Belfast Insane Asylum, 1887, HOS 28/1/2/5.

[26] Belfast and Ulster Directory and PRONI, Valuation Books, (VAL 12B/23/7e, p. 285).

daughter of the above Hugh Campbell, and wife of the late Capt. Samuel Lemon.[27]

Captain John Scott's assumption of command of the *Struan* could be fairly described as quick, decisive and thorough. From the *Mohur* he brought with him the new second mate, A. Cameron, from Prince Edward Island, Canada, employed at $35 per month, $10 more than the wage Pollack had been promised as second mate on signing the articles at Liverpool; three experienced able seamen from the *Mohur* crew, probably men known to be competent at sailmaking, to repair damage incurred while all the sails were set during the time the vessel was grounded near Cedar Key; and the *Mohur's* steward. Thomas Nagle, an able seaman from Waterford, Ireland, was discharged into the hospital immediately upon Scott's taking command. Whether Nagle was sick or had suffered an injury aboard the *Struan* has not been discovered.

Scott apparently accepted Pollack as first mate and designated the officer he brought from his previous command as second mate for the trip to Havre. Perhaps the reason Pollack did not progress to command of the *Struan* is because he did not have a captain's certificate. Samuel Lemon himself had sailed for several years as second or first mate aboard the *Struan* while holding a limited certificate which did not permit him to ship as a captain. The owners and underwriters could have been justified in their desire to place the vessel in charge of a more mature commander, particularly in view of the damage to the hull that might have occurred during the multiple strandings. John Scott was nearly 15 years older than Pollack and had been mate on or commanded more than a dozen different vessels during his career.[28]

We were not able to discover the myriad of telegraphic messages that were inevitably exchanged among the ship's agent at New Orleans (Silas Weeks & Co.), the owner in Liverpool (Andrew Gibson), and even the insurance underwriters, who must have been critical of Pollack for waiting too long to wrest control of the ship

---

[27] *Gravestone Inscriptions for Bangor*, edited by R.S.J. Clark. Bangor, County Down, Northern Ireland.

[28] John Scott was born at Richibucto, New Brunswick in 1842, and was issued Captain's certificate No. C 85046 after examination at Liverpool in 1870. His service began as mate on the *Juliet* in 1874, aboard the *Vancouver* in 1875-76, and on the *Grampian* in 1876-78. He was captain of the *Anglesia* for five years in the early 1880s, and then transferred to the *Mohur* in 1885, and to the *Struan* for 1887-88. He continued to command ships until well after the turn of the century, and lived until January 16, 1930.

from Lemon when the captain began to show signs of insanity. At just over thirty years of age, Pollack may have been considered too young or inexperienced for command of the ship.

Since Andrew Gibson owned both the *Struan* and the *Mohur* outright, it was relatively simple for him to instruct the New Orleans agent what he wanted done. By transferring Scott to the *Struan*, which was loaded ready for sea, he expedited departure of that ship while gaining a month to look for a new captain for the *Mohur*, which would be three or four weeks in loading.

The exact date of the *Struan*'s departure from Port Eads is not recorded, though Captain Scott specified in hiring the new crew members that they should be aboard ready for duty on the morning of March 3, 1887. It can be safely assumed that as a cautious and experienced skipper, he used several days to check over and inspect every piece of standing and running rigging, every sail and every spare piece of equipment to avoid surprises enroute to Havre, where he arrived on April 26, 1887.

The *Struan* was reported spoken March 15, 1887, but the location of the alleged sighting (26° N latitude, 100° W longitude), as specified in the April 22nd issue of Lloyds Weekly Shipping Index, is a dry land area in the vicinity of Monterrey, Mexico—well over one thousand miles from where the *Struan* would have been at the time. It is apparent from this incident that Lloyds officials either accepted ship positions as correct when received and made no effort to check their accuracy or that in their rush to publish up-to-date information they did not proofread the copy after it was typeset.

A later sighting on March 31 was reported as 46° N latitude, 38° W longitude, about a thousand nautical miles south of Greenland, and indicated that Captain Scott was attempting to utilize the natural flow of the North Atlantic drift currents to speed his vessel on its course.

It is significant to note that Captain Scott, after having observed Pollack's abilities and seamanship for almost two months on the voyage from Port Eads to Havre, did not offer him the first mate's berth on the *Struan*'s subsequent trip from Havre to South America, but hired another first officer over him. Pollack willingly signed on for that trip, however, again as second mate.

Chapter XIII

Ballast was a Dangerous Cargo!

Most dictionaries in discussing ballast provide the literal translation meaning "worthless load" from the Danish "ballast," and then define the term as "heavy matter such as stone, sand, or iron carried in the bottom of a ship or other vessel to prevent it from being readily overturned." A vessel is said to be "in ballast" when it sails without a cargo, *i.e.*, without a lading *carried for pay.*

Two hundred years ago, when vessels were much smaller than they are today, a few tons of local rock carefully placed near the keel of a three or four hundred ton vessel would insure its seaworthiness with a minimum of bracing, bulkheading or blocking to keep it in place. As the average size of newly constructed wooden vessels increased in the 1870s to 1,500 net register tons or more, ballasting became a more difficult chore. Buying and placement of substantial tonnages of rock or stone became cost factors to be carefully considered in vessel operation. Purchase of timber for bracing the ballast in place and labor expended for installation of that bulkheading, together with costs related to removal of the ballast and its bracing at the end of a voyage often resulted in a sizable expenditure, sometimes approximating a substantial fraction of what would have been spent to load a paying cargo.

As long as ships continued to be constructed of wood, ballast contractors' workmen and ships' crews, usually under supervision of the carpenter and the ship's officers, could nail blocking, braces, cleats and planking directly to wooden ribs and beams. With the advent of iron and steel vessels, the fastening of such bracing and blocking to or against iron or steel ribs or beams became more difficult, and required new skills and methods not known to many old-time ships' carpenters. Recognizing this problem, some builders constructed vessels using alternating ribs of wood and steel.[1]

To avoid shoaling of their harbor facilities through indiscriminate disposal of ballast from arriving vessels, most ports had regulations governing dumping of ballast, designating specific areas where such unwanted material could be discharged overboard. At Cardiff, Wales huge ballast cranes were erected at the Bute docks prior to the mid-nineteenth century. Vessels were allowed to dock at the inner harbor in ballast; the cranes transferred the ballast to stockpiles on dry land where it was made available to builders and townspeople who had a use for it. Even today, a tastefully constructed substantial building near the waterfront there incorporates major quantities of rock salvaged more than a hundred years ago from the ballast piles at the Bute docks.

One writer discussing the maritime history of the North Florida area describes in detail the disposition of the many varieties, shapes and sizes of ballast unloaded by the hundreds of vessels that visited Pensacola before the turn of the century to load the 200 million board feet of pine timber and lumber exported from that locality each year.

The city increased its waterfront area substantially by requiring that unwanted ballast be used to fill in low-lying coastal land. Rather than earning a reputation as an international dump site, the community took pride in having within its environs material from a dozen European and South American locations, as well as lava rock from Martinique.[2]

Ballast, as such, has been of substantial importance to the seafaring communities of all nations since very early times. Other than a simple flat-bottomed rowboat or scow, any vessel designed to carry cargo in a hull projecting below the waterline needs ballast for stabilizing the craft when no cargo is being carried. Some larger hulls were built with the intention of having them float on an even keel when unloaded; others were discovered to have that ability after they were launched. These vessels had a distinct economic advantage over cargo carriers that required ballast for stability even in protected harbors. Owners of vessels needing little or no ballast in anchorage-to-dock or dock-to-dock movement when under tow by a tug (with no danger of having to hoist the sails) would emphasize, "shifts easily without ballast" or "stands without ballast" when making a sale offer. Advertisements for several such vessels appeared in the July 2, 1877 issue of the *Liverpool Journal of Commerce* while the *Struan* was unloading there.

---

[1] See Chapter I, "The Clythia Marbles" for comments by the captain of the salvage vessel *Big Wig* relating to construction of the *Clythia*.

[2] Gloria Jahoda, *The Other Florida* (Port Salerno, Florida: Florida Classics Library, 1984), p. 37.

The bark *Tamar E. Marshall*, which rescued the *Struan* crew ten days after that vessel was dismasted and waterlogged in a series of violent storms, could not accurately be described as being able to "stand alone" without ballast.

The TEM was anchored out in Puget Sound near the dock at which she was scheduled to load sacked wheat at the same time the *Struan* was taking on a lumber cargo at Port Discovery in late 1890. The *TEM* was towed to the wheat dock to receive a partial load called a "stiffening" (transported across the deck of another vessel then loading its own cargo), before her captain felt comfortable in disposing of the last of his ballast out in the stream.

Some captains, especially when unloading cargo into lighters at an open anchorage, rigged logs, spare yards or even their own yards alongside their vessels for additional stability until the outbound cargo became available, or until unloading was complete and ballast could be brought aboard. It was especially important when depending on such rigging that all lashings on pontoons be tight and secure, with provision made by athwartship chains or cables secured under the keel to keep the floating stabilizers from "riding up" if the vessel began to tip.

By the beginning of the 19th century, shipmasters were being encouraged to exercise greater care in the selection, stowage and discharge of ballast materials. An article in the Naval Chronicle, Volume IV (1801) p. 270, entitled "Danger of Shingle Ballast"[3] (recently transcribed and placed on the Internet by Lars Bruzelious) tells the plight of the ship *Mysore*, an Indian vessel of substantial size and quality, which was almost lost at sea on a voyage from Bengal to Bombay because of carelessness in the stowage and removal of ballast by a previous crew on an earlier voyage. Some three weeks before arrival at Bombay, she sprang a leak so serious that the crew had to work at the pumps around the clock to save the ship from foundering.

Ballast

Ships requiring such from Dundee to Birkenhead or Liverpool can be quickly supplied with Craigie Stone on favourable terms upon applying to W. MADDOCKS & CO., Sole Agents, Corporation Road, Birkenhead, and 11 Castlestreet, Liverpool. DEALERS AND CONTRACTORS IN ALL KINDS OF SHIPS' BALLAST.

Advertisement appearing in the *Liverpool Shipping Telegraph and Daily Commercial Advertiser*, Saturday, July 3, 1886.

On arrival in Bombay, the ship was docked for inspection, where it was discovered that an almost-round stone, approximately four inches in diameter, from ballast used years before had slipped through a missing, damaged, or defective floor timber on the lowermost deck, where it had lodged against the keel, resting also on the inner surface of the garboard strake[4]. There, confined to a narrow pathway between the keel on one side and the vessel's hull on the other, and limited in its forward and backward motion by two of the vessel's ribs intersecting the keel, it ground its way through the garboard strake from the movement of the vessel over a long period of time. Even the keel itself was indented some three inches by the scouring action of the stone's constant travel. The damaged area was large enough that a workman was able to put his arm through the hole.

Most members of the maritime fraternity on learning of the *Mysore* problem would have faulted the vessel's officers for not carefully inspecting the deck on which the ballast was to be placed before it was placed, and double-faulted them for not noticing damage to the deck timbers when the ballast was removed and the hold cleaned.

To facilitate the loading and unloading of ballast, some cargo vessels were constructed with square ballast-ports located about midway between the waterline and the weather deck, allowing ballast to be put aboard without having to lift it up and over the main deck to the cargo hatches. When so equipped, a vessel could load or discharge ballast even in a driving rain without uncovering the main hatches or causing damage to the deck or to painted surfaces.

Some shipmasters, however, and many vessel owners, took a dim view of these installations. Captains who had reputations for fast outbound and inbound passages had little use for ballast-ports. They pointed out that there was seldom a need for handling ballast, and that the extra costs involved in opening, closing and maintaining extra hull openings offset any saving in the transfer of ballast. Skippers who habitually pressed a vessel to its limits often ran for hundreds of miles with the "lee rail awash." Under these conditions it would have been impossible when carrying a bulky lading like wool or cotton to con-

---

3 The term "shingle ballast" does not indicate, as might be inferred from the term itself, bundles of shingles used as ballast; but small, water-worn stones or pebbles such as lie in loose sheets or beds on a beach. (First reference in 1530-40). The term "shingle" in this connotation is believed to be a derivative of the Norwegian "singel," defined as coarse gravel or small stones.

4 The first exterior hull plank immediately adjacent to and on either side of the keel of a wooden vessel.

stantly monitor the ballast-ports, to detect leakage.[5]

Recognition of the danger to vessels carrying ballast was noted in an official publication of the British government prior to the turn of the century.[6]

A table enumerating maritime losses for a one-year period showed that more than twice as many vessels carrying ballast became total losses in strandings (running ashore and breaking up) as did those loaded with timber. Keeping in mind that timber cargoes were normally tightly packed to accommodate the greatest possible load suggests that shifting ballast might have contributed to losses of some of the ballast-laden vessels.

The table also demonstrates that three times as many vessels sailing "in ballast" turned up missing during the 12-month period as did those carrying timber cargoes. Perhaps some shipmasters, to avoid "unnecessary expense," carried less ballast than might really have been required, and as a consequence were left with higher-floating and less maneuverable craft in storms or other adverse conditions.

Another situation which may have had a bearing on the loss of at least some vessels over the years echoed the pernicious greed of a few sea captains, probably in most cases those who did not own shares in their ships. If, for instance, a master's records presented to the owner at the conclusion of a voyage reflected disbursement of $1,000 for 800 tons of ballast taken aboard after discharging the initial cargo, there was no satisfactory way for the shipowner to ascertain whether that full amount was, in fact, actually delivered and utilized. One author,[7] writing in 1997, alleges that it was "accepted practice" for a master to "line his pockets" by ordering and taking a bill of sale for a quantity of ballast material from a supplier who then, by previous arrangement, delivered an amount less than the invoice specified, after which the captain and the ballast merchant with whom he was in collusion would split the value of the unfurnished portion.

The British bark *Guiana*, bound for New York carrying a full cargo of sugar from the Philippines, put into St. Thomas, West Indies[8] March 18, 1886 with at least a foot of water in her hold from storm-induced hull leakage. Pilotage records indicate her entering draft was 21 feet. After unloading the entire cargo and doing some

---

[5] See Chapter IX, "Admeasurement, Tonnage and Cargo" for a discussion of fires caused by moisture in wool cargoes.

[6] *Papers of the House of Commons*, Part III, Table 22, page 27.

[7] Richard Woodman, *The History of the Ship* (London: Conway Maritime Press, Brassey's UK Ltd, 1997) p. 208. Also the Lyons Press, New York, N.Y.

minor temporary patching, the ship's agents purchased a little over 500 tons of stone ballast for 70¢ per ton, and paid a local contractor $110.50 to put it on board and "trim the load." However, no record of any payment to a contractor or work group is reflected in the meticulously kept records for the vessel, suggesting that the ship's crew bulkheaded and made secure the ballast, or that the work was not done at all.[9]

On leaving St. Thomas, the draft of the Guiana was only 12 feet. Allowing a foot of draft for the water in the hold on arrival and loaded to only one-fifth capacity, the bark left eight feet higher in the water than she would have been if carrying a full capacity cargo.

In this particular instance, Captain Eben C. Byrns purposely chose to "run light" to keep as much as possible of his damaged hull planking above the waterline, thereby easing the load on the pumps. He reached Saint John, N.B. safely for permanent repairs some three weeks after departure. (See Chapter XV). Captain Byrns took a calculated risk in sailing such a lightly ballasted vessel up the East Coast of the United States just before the hurricane season—but he may have reasoned that he could reach harbor in New Brunswick by early May, before the storms that intermittently devastate the area during summer months began.

"Hogging" is the gradual deformation of the keel of a wooden vessel resulting from the upward pressure of the water under the vessel (which equals the downward pressure on the water from the combined weight of the ship's hull and the cargo or ballast carried).

This deformation, resembling the arch of a hog's back, begins at launching and gradually worsens during a ship's life. It may be inconsequential early in a vessel's service, and less rapid in its progression while a hull is unloaded or lightly loaded, but nevertheless continues

---

[8] now the U.S. Virgin Islands

[9] When ballast was required at a major port, it could usually be obtained from a dealer in that commodity. Most such dealers would also contract to place and stabilize the ballast if requested to do so, and were paid an additional sum for that work. Some captains, however, required their crews to put the ballast in place after it had been brought alongside or aboard. At small or little-used ports, where ballast dealers had not been established, it sometimes became necessary for sailors to bring ballast materials to their vessel in small boats. Activity of that nature is believed to have been the basis for an old sailors' ditty:

> "Let the captain find a cargo
> For Hamburg or Key Largo
> So our hands will not be calloused
> By handling of the ballast."
>
> —Anonymous.

every day that a vessel remains afloat.

Hogging negatively affects a vessel's speed and maneuverability. The concave cavity under the keel interrupts the flow of water past the hull as the ship moves forward, limiting its speed. With pronounced hogging, more water is in effect trapped, further decreasing the speed.

In considering the causes of hogging, it should be remembered that most cargoes, because of their bulk or weight, were loaded into hatches in the midship areas, rather than in the bow and stern sections, because sharp corners and confining dimensions sharply limited useable space in those locations.

In a wooden vessel, the ribs near the bow and stern, being relatively short, were formed from a single piece and acted almost like posts reaching from the keel to the top deck, under little pressure to bend. Ribs at the mid-section of the vessel, however, were in many cases two or three times longer from keel to weather deck than bow and stern ribs and were usually "built up" or formed from several pieces fastened together to produce a single rib.

Concentrating the cargo in the midship section of the vessel put a strain on the ribs in that area, gradually exerting a bending pressure on the ribs' joints as well as on their individual timber sections, allowing the keel to rise at mid-vessel as the ribs attached to it lost their original shape from the crushing pressure. In older vessels, ribs have been known to bend to the point of collapse, allowing the keel to be forced upward amidships, fracture, and break the ship apart.

Lumber cargoes, due to requirements for straight-line stowage involving long timbers, were perhaps the greatest contributors to hogging of a vessel. Heavy loading in the midship section of the hull, and the extra weight of a substantial deck load in the same area could lead to crushing pressure on the midship ribs, leading to even greater hogging.

The danger of hogging in wooden vessels was recognized and successfully dealt with in construction of the USS *Constitution* and five other wooden frigates built in 1797 for the fledgling American navy.[10]

Considering the fact that shipowners preferred not to transport timber cargoes in newer wooden vessels because of frequent interior hull damage during loading and unloading the heavy timbers through bow and stern ports, and the "hogging" action on the hull itself that inevitably resulted after a couple of timber cargoes, it could be assumed

---

[10] Roger Archibald, "Six Ships That Shook the World," American Heritage of Invention and Technology, Volume 13, Number 2 (Fall 1997) pp. 28-30.

that some of the vessels lost with only ballast in their holds were of newer construction. One such vessel has been carefully researched.

The *Senegal*, vessel No. 97865, signal letters MGPT, was a riveted steel sailing bark of 1,352 net register tons, built by Russell & Co. at Port Glasgow, Scotland, in 1891. Newspaper accounts of the launching described the vessel as 238 feet long, 36 feet in breadth, and 21 feet, 9 inches in depth. Its construction was ordered by Gilbert M. Steeves, a Liverpool shipowner. After the hull was in the water, it was towed to a Port Glasgow dock for installation of masts and rigging and completion of interior fittings.

Wallace identifies Gilbert M. Steeves as a Saint John, New Brunswick resident when in 1874 he ordered construction of the 1,274-ton wooden ship *Alexandrovna* by a yard in Rothesay (now a suburb of Saint John).[11]

Steeves formed a limited partnership—The Ship *Senegal* Company—with other members of his family to hold ownership of the vessel. The first commander was Captain I. Webster, soon followed by Captain W.A. Steeves, believed to have been either the brother or son of Gilbert M. Steeves.

Completing its first voyage, it left Calcutta, India on February 23, 1893 and was spoken twice during its trip back to Britain—once on May 18 in latitude 13°S, longitude 8°W, and again on July 16th about 20 miles east of Cape Wrath.[12]

The *Senegal* arrived at Dundee, Scotland July 22 and sailed from there for Swansea on August 7. Swansea at the time was one of four ports in Wales—Cardiff, Newport, Swansea and Barry—well known for the export of coal. There was a change of masters at Dundee; Captain W.A. Steeves took over from Captain I. Webster. The *Senegal* arrived at San Diego, California on February 14, 1894. The cargo was unloaded and the ship took on ballast for a trip to Puget Sound, intending to load either lumber or wheat in the Tacoma area for

---

[11] Frederick William Wallace, *Wooden Ships and Iron Men* (Boston, Charles E. Lauriat Co., 1937) p. 250.

[12] Cape Wrath is on the extreme Northwest point of Scotland, indicating that the *Senegal* chose a route north of the Scottish mainland to approach Dundee from the North. Such a voyage would have required that the captain elected either to take his ship through the Pentland Firth, a six- to eight-mile wide channel between the Atlantic Ocean and the North Sea, separating the mainland of Scotland from the Orkney Islands, or northward around the Orkneys to avoid the fast-moving tidal currents and dangerous whirlpools which made navigation difficult in the Pentland Firth, particularly for sailing vessels.
American Peoples Encyclopedia (Chicago: The Spencer Press Inc., 1952) 15-482.

Queenstown, Ireland, now known as Cobh. After sailing from San Diego on March 15, 1894, the *Senegal* never reached its next port of call (or any other port) and was posted as missing by Lloyds on July 20, 1894.

C. Olsen, a sailor on the schooner *Mary Cruss*, reported shortly after the *Senegal* was posted overdue that his vessel had been in company with the *Senegal* on the night of March 18, 1894 during a storm that he described as "the worst he had seen in twenty years on the ocean." The two vessels, he said, fell in together just as darkness was closing in. The *Senegal* was "beating up against a freshening northeast gale."

Olsen described how he watched for the *Senegal's* red light (which would have been the port running light) to appear, disappear and then reappear as the two vessels were alternately riding crests of waves or wallowing in troughs of the seas.

By midnight, the storm was at its worst, and he continued to watch for the intermittent reappearance of the *Senegal's* light until almost dawn, when that vessel seemed to rise on the crest of a wave, just in time to be attacked by a squall even more vicious than before. The red light gleamed, disappeared, reappeared momentarily, and then went out permanently. When dawn broke an hour later the *Senegal* was nowhere to be seen.

Olsen's report, though circulated widely among seafarers, received little credence in the press until Captain Thompson of the *Mary Cruss* wrote from San Diego three months after the incident to verify Olsen's description of the night's events as accurate, and to confirm latitude 30.17 N, longitude 119.08 W as the approximate location of the *Senegal's* disappearance.[13]

The same newspaper clipping relates that the *Senegal* left San Diego March 15 with "loose sand as ballast." If that information is correct, the captain selected probably the most dangerous and most difficult to contain type of ballast available. Loose sand, by its very nature, particularly when dry, seeks out the tiniest cracks and crevices in the bulkheading built to confine it—leaving spaces for more sand to fill when a ship is being tossed in a storm. Wooden barriers built to confine the ballast and keep the vessel on an even keel in port, could easily break loose from any major movement of the mass, permitting serious shifting of ballast and endangering the vessel's recovery from an extreme roll.

---

[13] The preceding four paragraphs are paraphrased from a yellowed newspaper clipping believed to be from the Portland Oregonian of about September 21, 1894.

An ironic twist to this story is that in 1893, *before* the *Senegal* was lost, the same shipyard that built it turned out another steel bark, the *Bermuda*, with built-in water ballast tanks and a steam pump to adjust their water levels. In just two years, contemporary ship design had progressed from loose ballast in the hull to seawater confined in a tank.

The idea of ballast tanks to hold seawater was not immediately endorsed by some shipowners. Although the extra steel for construction of the tanks and the steam pump, piping, and other equipment required to operate them added, perhaps appreciably, to the initial cost of the hull, it took some time for owners to appreciate the fact that ballast costs would be eliminated for the life of the ship.

No one foresaw that various forms of small marine life could be sucked into a vessel with its water ballast and be routinely discharged at ports half-way around the world, eventually resulting in these unwelcome guests being introduced to areas previously free of such contamination.

When sand ballast in Rio cost 7s 6d a ton, a saving equal to about $2,000 was available to a 1,500-ton vessel merely by opening the valve and taking on free seawater each time ballast was needed. In addition, the new method required much less time than placing solid ballast and bracing it in place. As an added advantage, captains who had been prone to take aboard too little ballast for safety under extreme conditions were motivated to keep the tank full , since no expense or delay was involved..

Ballast was considered by mariners as an unwelcome necessity that had to be purchased, loaded, stowed, transported and later unloaded and abandoned, usually without a financial benefit to the vessel. Some shipmasters therefore, trivialized the necessity for exercising the same degree of care in its stowage that they would automatically have given to the handling of "for hire" cargo.

Safely preparing for sailing with sand ballast at the time of the *Senegal's* loss would have entailed one of two possible methods—either of which was also recommended as a means for confining bulk wheat cargoes. If each steel rib of the vessel had been equipped with an eye bolt or clevis near the lower-deck level, steel cables or chains and turnbuckles could have been connected to the rib fittings, and the loose sand "crowned" in the center of the vessel. Three layers of carefully secured sacked sand would then have been laid in an interlocking pattern to cover the cracks in the next lower layer. Over this, two-inch or heavier wooden dunnage strips would have been laid full length parallel to the ship's keel, about three feet apart to allow the athwartship cables or chains to cinch down the entire mass, confining the ballast even under the most severe conditions.

If eye bolts, clevises and cables/chains and turnbuckles were not available, the other recommended method of stabilizing the ballast would have involved loading the loose sand evenly across the holds, similarly covering it with sacked sand and longitudinal dunnage strips, but adding an additional layer of well-nailed heavy dunnage "athwartship" (at right angles to that laid parallel to the keel). Four-by-six inch bracing could be then placed at five- or six-foot intervals from the top layer of dunnage up to the underside of the steel deck beams. That method was a more time consuming operation than the first, since properly fastening heavy wooden bracing to the underside of steel beams required extra labor as well as a great deal more lumber, nails, bracing and blocking than might have been necessary in a wooden vessel.

Certain types of ballast presented fewer problems. Rough, irregular-shaped rocks could be partially interlocked in their placement, to prevent their movement at sea. Smooth stones needed more protection from "wandering." Brick-like "cobble stone," often selected as ballast for ships headed to cities like Portland, Oregon, where they could be sold for street-paving, could be neatly packed together, but even if placed haphazardly, they were fairly secure from movement.

Today, more than one hundred years after the *Senegal's* loss, nobody knows for certain why the vessel went down. Only three years old, built by a leading Scottish shipbuilder to the most exacting specifications, the bark could reasonably have been expected to serve its owners well for 25 or 30 years. Many such ships did just that.

The fact that the *Senegal*, on the night it sank, was in the company of a much older and substantially smaller schooner which weathered the storm without damage, raises questions about what mishap befell the bark. Newspaper reports after the sinking referred to the fact that the captain had brought his young daughter on the voyage. It could be reasonably inferred that he would have made extra effort in the operation of his vessel to insure her safety.

Writing in 1929, Lubbock alluded to rumours at the time of the *Senegal's* loss that the vessel may not have been adequately manned. His suggestion that the entire ship's company on her final voyage consisted of the master, two mates, three apprentices, steward and stewardess, carpenter, six A.B.s and two ordinary seamen, gives the impression that he was quoting some knowledgeable authority (as, for instance, British consular records), though he fails to identify his source.[14]

It should be noted, however, that a complement of 17 aboard a

[14] Basil Lubbock, *The Last of the Windjammers Vol. II* (Glasgow, Scotland: Brown, Son & Ferguson, Ltd., 1929) p. 209.

John Henry (Harry) Heggie, an apprentice aboard the 1,352-ton British riveted steel sailing bark *Senegal*, Vessel No. 97865, lost in a storm the night of March 18-19, 1894 in prox. latitude 30.17 N, longitude 119.08 W, while enroute in ballast from San Diego to Puget Sound.

bark of the *Senegal's* size was within general practice. Although British vessels routinely carried more apprentices and somewhat larger crews, many full-rigged ships were being, and had been for many years, operated by the thrifty Norwegians with only 18 persons aboard. Assuming that the apprentices were experienced enough at making and taking in sail, and keeping in mind that a bark had one less set of square-rigged sails than a ship, a total crew of 17 should have been able to handle the vessel safely.

My own interest in the *Senegal's* fate stems from the fact that my only uncle on my mother's side of the family, John Henry (Harry) Heggie, a 19-year old apprentice aboard the *Senegal*, was lost with other members of the crew and the captain's daughter. May they all rest in peace.

Chapter XIV

Coal Out, Nitrate Back—Coal Out, Maize Back

After unloading at Havre the cotton and staves brought from New Orleans, Captain John Scott (skipper of the *Struan* since February 27, 1887), hired a minimum by-the-run crew on May 18 to take the vessel to Cardiff, Wales for loading coal destined for South America. He was apparently pleased with that crew's performance on the six-day passage and demonstrated his pleasure by endorsing on the ship's articles a "very good" rating, both as to ability and general conduct, for every person aboard.

The mate, Joseph W. Morrell, listed his birthplace when he signed for the trip as Yarmouth, Nova Scotia. A Dutch citizen, Louis Couburier, held the steward's position, but no cook appeared on the roster. Ten able seamen rounded out the crew. By happenstance, four of these had just served aboard the bark *Lynwood*, built in the same yard as the *Struan* by John Fraser in 1880.

On arrival in the Cardiff-Penarth area May 24, the crew were paid off, leaving the captain in sole possession of the ship during loading, which in this instance required only two weeks. On June 8, 1887, contemplating an extended voyage to South America, Captain Scott engaged a full crew, including Joseph Morrell again as mate, but on this trip that officer gave his birthplace as Brooklyn, New York rather than Yarmouth, Nova Scotia. Samuel Pollock, previously aboard the *Struan* as second mate under the deposed Captain Samuel Lemon, came again as second officer. Murdo Gunn from the Isle of Lewes filled the carpenter's post; Henry Monroe was hired as cook and steward two days later after the person originally scheduled for that billet failed to appear at the appointed time. Fifteen ABs, two ordinary seamen and a 15-year old apprentice boy fresh off the training ship *Endeavor* completed the crew.

Scott's wisdom in hiring a full complement of sailors was confirmed when three of the ABs hired did not join the ship by sailing time; they were not replaced. The *Struan* cleared from Cardiff on June 9, 1887 with 2,472 tons of coal, and was spoken July 12th southwest of Cape Verde Islands, at 10° N, 27° W, about halfway to Montevideo, where she arrived August 16, ten days after the sighting was announced in the Lloyd's Weekly Shipping Index.

Captain Scott and combination cook/steward Henry W. Monroe agreed on September 13 "by mutual consent" that Monroe should leave the ship. Although three able seamen were hired in Montevideo at that time, the master couldn't find a suitable replacement for the cook/steward until October 1, when Portuguese citizen G. R. Soaris was employed. Two seamen were hospitalized with sickness and their accumulated pay left with the consulate.

Louis Oliver, an AB from France, died November 10 aboard the *Struan* while the ship was enroute from Montevideo to Valparaiso, Chile, where the vessel arrived on November 24, 1887. Scott reported Oliver's death to the vice-consul at Valparaiso, and upon exhibiting the entry in his official logbook relating to the seaman's demise, obtained an entry in the articles acknowledging report of the death, and stating that an "order on the owners" for £6 18s 6d due the deceased, and the deceased's unsold effects consisting of used clothing, had been delivered to the consulate. No details have survived concerning Oliver's death, not even a reference either to sickness or accident. In connection with investigation of other deaths aboard the *Struan*, it was the usual practice for the consular official conducting the initial inquiry to endorse the articles to show "that no blame could attach to any person aboard the ship." Whether or not the lack of such phraseology could be considered significant in this instance is a matter of conjecture.

The *Struan* didn't leave Valparaiso until February 8, 1888, at which time the vessel departed for Pisagua, Chile, dropping anchor there on February 17.

In addition to the previously mentioned difficulty in sailing northbound along the West coast of South America because south winds and the northward Humboldt current cooperated to drive a vessel past her intended destination, other dangers presented themselves to shipmasters negotiating that route. Earthquakes and the resulting tidal waves had wrecked many Chilean ports several times in the 19th century, with stricken vessels sometimes being hurled ashore far beyond the normal high tide line by sixty-foot breakers. Another hazard involved what were called the "northers," fierce winds which sometimes swept down to create havoc among vessels anchored in open water near and adjacent to various ports. When a crowded anchorage, such as commonly occurred at Iquique, Pisagua or Valparaiso was struck by such a wind, some vessels inevitably broke their cables or dragged their anchors, causing considerable collision damage to themselves and the other ships near them.

Pisagua in those days was primarily a port for loading one commodity—nitrate—variously called nitrate of soda, Chile saltpeter, sodium nitrate ($NaNO^3$), (in Great Britain nitre) or in the maritime community, just plain "soda." It was mined in the great salt desert of Tarapaca, and in earlier days had been brought to the port on the backs of pack mules and donkeys and transferred by lighters propelled by two-man crews to sailing vessels anchored offshore in an open roadstead. By the time of the *Struan*'s arrival, donkeys had yielded their chores to a primitive railroad, but the same hand-propelled lighters transported the nitrate to the waiting ships, greatly slowing the loading process. Iquique and Pisagua depended for the continued well-being of their nitrate exporting business on the Desert of Tarapaca and the rich deposits of the resource found in that location, which had been worked for half a century.[1]

Nitrate aboard ship caused their captains many problems. It was very deliquescent, meaning that it had a propensity to absorb moisture from the air and melt into liquid. It needed to be completely dry when stowed aboard, and required more than the normal amounts of dunnage. The greatest risk in handling it was the extreme danger of fire. Knowledgeable shipmasters when loading nitrate kept barrels of water near each hatch and at other convenient locations, not just ordinary water, but water which had been pretreated by soaking nitrate in it, reputedly the only mixture that would extinguish burning nitrate. One author lists from memory three ships destroyed by fire during nitrate loading operations.[2]

For some time prior to 1900, Chilean copper ore, guano, and nitrate were cargoes shunned by first-class ships. No shipmaster in his right mind would have loaded nitrate into a second-class vessel since even a small quantity of salt water in the cargo hold could have precipitated a hazardous and expensive mess. As more European farmers began to use "Chile saltpeter" for fertilizer and other uses were found

---

[1] Isaiah Bowman, *Desert Trails of Atacama* (New York: American Geographical Society, 1924) pp. 62-63.

[2] Basil Lubbock, *The Nitrate Clippers* (Glasgow, Scotland: Brown, Son & Ferguson, Ltd., Nautical Publishers, 1932) pp. 5-6.

for the nitrate, the market expanded and more and more vessels were willing to carry it. In 1888, the year the *Struan* loaded at Pisagua, it has been estimated that 300,000 tons of nitrate were exported from that area of Chile. Such a quantity would translate to more than 180 full ladings for vessels of the *Struan*'s size—one shipload every other day of the year.

With the last of the nitrate cargo finally taken aboard without serious incident, Captain Scott abandoned his anchorage at Pisagua on April 8, 1888 and departed, specifying for authorities Hampton Roads, Virginia, U.S.A as his intended destination. Either that information was given to conceal the real intended port of discharge, or he changed his mind enroute, since the *Struan* showed up in New York harbor on June 28.

> The British ship *Struan* arrived June 29, 1888 at New York (Scott) from Pisagua 81 days enroute, with soda to order, vessel to master.[3]

A dozen crew members were discharged at New York. The first of these, the cook/steward Soaris engaged at Montevideo, left only three days after arrival there. When Captain Scott hired a replacement just 48 hours prior to sailing, he must have realized that by the time his vessel reached its home port at Liverpool he would have gone through four combination cook/stewards on a single voyage of less than 15 months. On a subsequent trip, he chose to engage separate persons for the two jobs, and each of them stayed with the ship for the entire 13-month cruise. From an expense standpoint, breaking the jobs down to be held by two different persons cost very little more than the combined service—less than the equivalent of U.S. $10 per month.

The *New York Tribune* published a special warning/notice to mariners two weeks after the *Struan*'s arrival, pointing out that July and August were the most dangerous months for hurricanes along the East Coast, and noting that many yachts and other smaller craft had recently been cast ashore in the Boston area by such winds.

In its July 14 issue, the *Tribune* publicized as a news item a story of the invention of rat guards, recalling for its readers that:

> the skill with which dock rats board a vessel by running along her cables has always been a terror to shipowners. A protection has just been invented in the shape of a big tin funnel through which the

[3] The *New York Tribune*, June 30, 1888.

> cable runs. The big end of the funnel faces the shore, and the rats cannot surmount it.

Although use of some version of the funnel-shaped rat guards described above is required by law at most ports throughout the world, the expediency of that use is widely questioned. Dock and health authorities today are much more concerned about foreign rat populations coming ashore from moored vessels than the ship's masters were about rats boarding from shore in the *Struan*'s time. Because metal rat guards today are a modified blend of the funnel idea designed to calm the fears both of the dock keepers and the shipmasters, they are generally considered to be of little practical value from either standpoint, and are accordingly often carelessly or improperly installed.

The most-used device is a circular sheet metal guard some 30 or 36 inches in diameter, having a sliding panel to permit the device to be clamped around a mooring cable. Obvious weaknesses in the system devolve from the fact that the center hole must be made large enough to accommodate a three-inch manila hawser; when a 3/4" steel cable is used instead, however, there is left enough space around the cable to enable a rat of moderate size to get through without much difficulty. In many instances, the devices do not stay perpendicular to the mooring line, and sag to an almost horizontal position as cables become slack with tidal changes, permitting rats to walk across them without effort.

Detractors of the rat guard phobia point out that vermin as well as people can walk or run up gangways, that they can jump from dock to ship, or vice-versa, with little difficulty when the vessel is moored closely alongside the pier, or that they can burrow into a sack of cargo and find themselves carefully hoisted aboard or ashore by longshoremen.

One delivery made during the *Struan*'s stay at New York might be of primary interest only to naval history buffs. "About sixty tons of steel plate," the Tribune announced on July 14, "for use in construction of the new twin-screw battleship *Maine* have arrived at the Brooklyn navy yard from Pittsburgh."

> Sailed, August 10, 1888, the ship *Struan* (Br) (Scott) for Liverpool (Agents Pendleton, Carver and Nichols)[4].

Upon arriving in Liverpool on August 31, Captain Scott delayed engaging a crew until September 19 for what would be the *Struan*'s

[4] The *New York Tribune*, August 11, 1888.

final voyage under the British flag. After a three-day run to Cardiff in ballast, she again loaded coal at the East Bute dock and set sail for Rio de Janeiro on October 3, 1888. She was spoken November 4 in 2° N latitude, 30° W longitude (about six hundred miles northeast of Recife, Brazil) and actually reached Rio a few days before the news of that speaking was published in Lloyd's Weekly Shipping Index on November 23, 1888.

After a comparatively uneventful and reasonably speedy 45-day passage, the *Struan* arrived at Rio de Janeiro on November 17, 1888, carrying 2,452 tons of Wales coal from mines in the hills behind Cardiff.

Publication of vessel arrivals the next day included the entry "from Cardiff, 45 days, British full rigged ship *Struan*, 1473 tons, Captain J. Scott, crew 20, cargo—coal to Norton Megam & Co."

The vessel was "gauged" (inspected by customs and port authorities) on the 19th and directed to the "Ponta da Area" section of the port at Niteroi City for unloading.[5]

No mention is made in newspaper accounts of that time as to whether the *Struan* unloaded its coal to barges or lighters while anchored out in the stream, or whether the ship was brought alongside a dock to transfer its cargo. The ship's articles reveal that six men deserted at Rio. On a previous voyage to the same port two years earlier (under Captain Samuel Lemon) at which time the ship anchored out in the harbor and transferred its cargo to lighters to be ferried ashore, no crew members had deserted.

Cleared at Rio and ready for sea on December 12, 1888, the ship actually sailed, ballasted with stone, a few days later in time to arrive at Calcutta on March 5, 1889. Cargo loading at that port consumed only one month's time, but during that period many changes were made to the crew roster. Almost immediately upon the vessel's arrival there, three seamen deserted and were promptly caught and sentenced to six weeks in gaol. A few weeks later, another sailor went to gaol for a similar infraction, and a sick seaman was left in the local hospital.

From a study of ship's articles, both for the *Struan* and other vessels of the same time frame, it appears that deserters, at least from English-flagged vessels, were much more apt to be jailed for their absences at British colonial ports or in British controlled areas than if the same offense occurred in a foreign country. Rio de Janeiro and

---

5 "Jornal do Commercio" (RJ) - Microfilms: CPR-SPR 1 (162) at the Biblioteca Nacional, Avenida Rio Branco 219, Rio de Janeiro, Brazil. (Translated from the original Portuguese).

other South American coastal cities held a fascination for seafarers that enticed many to desert their vessels, but any effort on the part of civilian authorities at those locations to control or punish such desertion has not come to my attention.

An issue of the *Englishman's Overland Mail* being sold while the *Struan* was at Calcutta made note of the fact that the *Struan* was lying in the Hughli (Hooghly) River, erroneously tabulated her tonnage as 1,173, and listed her agents as Turner, Morreson and Co. Patrons of that publication were apparently more interested in the names of arriving passengers, horse racing results, and information about which lieutenants were on leave or being transferred, than on mundane commercial matters. Accordingly, no cargo information was included in the news items.

Some of the stories to which the publication gave its attention during the *Struan*'s visit included the fact that a Mr. Spencer was making balloon ascents at "remote locations" near Calcutta, and that he "surmised he had approached 13,000 feet in altitude." The mystery of why a visiting Chinese, one Am Lan, was buying up quantities of 2-anna coins was solved when the *Mail* explained to its readers that use of the coins was cheaper than purchasing buttons for use in the buyer's tailoring business.

The first mate, John J.E. Dean, from Liverpool, took his discharge at Calcutta, as did the German-born boatswain (bos'n) Charles Zerk. Captain Scott hired George Dick of Arbroath, Scotland as mate the next day and Ed Ablott of Boston as bos'n at the same time. Ablott was discharged on arrival in New York, and was not replaced. Forty-year old George Richards, from Swansea, Wales took the mate's job on September 20th at New York.

> The *Struan* sailed April 5, 1889 from Calcutta, was spoken July 8th off St. Helena, and arrived in New York August 21st.
> "Arrived August 22, 1889, ship *Struan* (Br) Scott, with maize, from Calcutta (136 days) to Raill Bros."[6]

Factual determination of what this cargo actually was could prove both difficult and confusing. Although the term "maize" in the United States is readily associated with "Indian corn," later shortened to "corn," different definitions apply in other countries. In England, the word "corn" commonly refers to wheat; "cornflowers" are so-called

---

[6] *The New York Tribune*, August 23, 1889.

because they are common weeds in wheatfields. In Scotland, a corn field is a field of oats.[7]

Taking into consideration that unlike corn, wheat is and was listed as a common export from Calcutta,[8] and that the cargo was being transported by a British owned and flagged vessel by a captain from Liverpool, it is reasonable to assume that the description "maize" referred to wheat.

On August 27, 1889, while the *Struan* was lying at Robinson Shores, Brooklyn, New York, 26-year old Norwegian seaman S.O. Gorgensen, while sitting on a stage painting the hull outside the starboard bow, was apparently seized with a fit and fell into the water and was drowned. British vice-consul Fraser at that port on September 23 made an endorsement on page 12 of the ship's articles, stating that he had inquired into the circumstances connected with the death of the said seaman and "find them to be as above stated, that proper entries have been made in the official log book, and that his wages and effects are both in charge of the master."

Having to offer $25 a month to get able seamen at New York to take the *Struan* back to Liverpool must have seriously offended Captain John Scott's frugal nature. The figure is exactly twice what such sailors were paid in British pounds when the ship left Cardiff 14 months previously, and almost twice the amount being paid in South American ports where the *Struan* touched. Unless he was completely dissatisfied with the performance of his crew, it is hard to understand why he would accede to the wholesale discharge "by mutual consent" of 14 seamen who had signed aboard at various ports for the duration of the present voyage, particularly when he faced replacing them at twice the wages they had been promised when hired. Two of these costly replacements were individuals who had just signed off another of the *Struan*'s sister ships, the 1,351-ton bark *Levuka* built by John Fraser at Saint John for his own account in 1884, and registered in Windsor, Nova Scotia.

The *Struan* sailed from New York October 4, 1889, and reached Liverpool on October 21, where the entire crew was discharged in preparation for turning the ship over to Norwegian purchasers.

---

7 *The American Peoples Encyclopedia*, (Chicago: The Spencer Press, Inc., 1952) 6-276.

8 *The American Peoples Encyclopedia*, (Chicago: The Spencer Press, Inc., 1952) 4-575.

Chapter XV

Insure or Not Insure?

Captain Skogland's poignant observation—"I wish we had insured" —in his written report in early 1891 to the owners in Norway about the loss of the *Struan*, did not reflect any omission of duty on his part, or failure by the managing owner to obtain insurance on the vessel.

Following the example of many Norwegian shipowners of the period, a conscientious decision not to insure, or to "go bare," had been made prior to the vessel's first departure from port.

Many Norwegian owners of sailing vessels at the time had accumulated fleets of substantial size, and considered the loss of a single vessel in a year less than the cost of marine insurance premiums on the entire fleet for a similar period. Another factor that should be kept in mind was that under Norwegian practice it was customary to divide the ownership of a vessel into 100 parts rather than the 64 specified by British law. It was therefore possible, if the operators chose to do so, to divide the ownership more widely and thereby dilute the risk of each participant in proportion to his or her purchased share. Rather than outright ownership of fewer vessels, some shipowners preferred to hold minor fractional shares in a larger number of vessels, feeling that this arrangement, in effect, limited their loss in event of a tragedy to a minimal amount in any particular ship. The other side of the coin, of course, was that there were more vessels represented in their pool of holdings and therefore a greater possibility of harm befalling any one vessel in the longer list.

Although their own shipbuilders constructed some very fine vessels, some of which remained in service for longer than the average 15 to 20 years usually recognized, the preference of many Norwegian shipowners of that time was toward purchase of 10- to 12-year old vessels from other countries, particularly Canada and England. These ships, having been substantially depreciated by their original owners, could usually be bought for about half the cost of new ones, and for some purposes served almost as well.

One well-known shipbuilder (skipsbygger) in Haugesund was John Hauge, winner of a gold medal for his ship models and designs at the 1878 Paris Exposition. One of his first vessels was the bark *Augvald*, owned by Arne Lothe (one of the owners of the *Struan*) after 1872. In 1876 he finished the full-rigger *Haugesund* for operation in partnership with Arne Lothe. Other vessels constructed at the Hauge shipyard (skipsverft) included the full-rigged *Zippora*, owned by H. O. Sundfor (another member of the *Struan* syndicate), the bark *Vikar*, and perhaps the most famous of all his vessels, the full-rigger *Oregon*.

Older vessels upon transfer to Norwegian registry were often fitted out with a wind-driven pump to keep the bilges dry. Some seafarers of that time professed to be able to identify a Norwegian vessel by the windmill contraption visible from a distance above the main deck even before that country's flag could be recognized. These pumps were sometimes jokingly referred to as "Scandahoovian House Flags" by the world's maritime community and some writers of the time. There was more method than madness to the installations, however, since the existence of such a pumping system precluded the necessity of turning out the crew for almost constant work at the pumps if a seam opened in a storm or some other misadventure caused a vessel to take on water at a rate threatening its safety.

Contrary to the usual Norwegian practice, the newly purchased *Struan* was not fitted out with a windmill pump while being initially prepared for sea at Liverpool in the fall of 1889. When Captain Skogland brought the ship into Plymouth, England only two months later for substantial storm damage repairs, a windmill pump was installed there as a part of the work on the vessel at that time.

Marine insurance has long been a subject little understood by the general population and perhaps somewhat misunderstood by those in the maritime community who might be expected to have the most concern for its scope and complexity.

General Average is a marine insurance term relating to the theory that all owners of a ship, cargo, and freight should be assessed a share of any loss by one or more of their number as a result of sacrifices made voluntarily of a portion of the ship or cargo in an effort calculated to save the residue and the lives of those on board from an impending peril. Common examples of activities subject to general average in the time of the *Struan* (if a vessel and its cargo were insured) would include a captain's decision to open the hatches and remove to be jettisoned a portion of stowed cargo to lighten ship in an attempt to refloat a stranded vessel. Similarly, cutting away masts or rigging in an attempt to right a badly listing vessel would qualify.[1]

The question of general average also arises when, in face of danger, sacrifices are made by the master of a vessel for the common good, or when damage is caused by water in extinguishing a fire on board ship. Statements are prepared apportioning the general average charges pro rata on the vessel itself, its freight earnings on that voyage and the value of the cargo. The amount so apportioned against a shipment of merchandise is collectible in full under the usual policy of marine insurance, providing the amount of the policy is at least equal to the arrival value of the cargo, or, if less than the arrival value, the underwriters are responsible only for the proportion of actual charges that the amount of insurance bears to the arrival value.[2]

Particular Average, on the other hand, is a loss to be borne entirely by the party whose property is involved or affected, toward which various parties must contribute. Many types of cargoes can be damaged by sea water from a leaking hull in a storm or minor grounding, and claims under a marine policy could arise from failure to deliver a portion of the cargo, as when machinery carried on deck breaks its lashings and goes over the side in a heavy sea.

A qualifying clause in common use in marine insurance certificates (in the late 19th and early 20th centuries) was known as the F P A E C clause, meaning "free of particular average English conditions." The wording was as follows:

> Warranted free of particular average unless the vessel or craft be stranded, sunk or burnt, but notwithstanding this warranty the assurers are to pay the insured value of any package or packages which may be totally lost in loading, transshipment or discharge, also any loss or damage to the interest insured which may reasonably be attributed to fire, collision or contact of the vessel and/or craft and/or conveyance with any external substance (ice included) other than water, or to discharge of cargo at a port of distress, also to pay landing, warehousing, forwarding and special charges if incurred, for which underwriters would be liable under a policy covering particular average.[3]

---

[1] In his written report to the owners after the *Struan* had been lost, Captain T. H. Skogland stated that he had ordered masts cut away in an effort to right the almost-capsized vessel. *See* page 174.

[2] Irving National Bank, *Trading With the Far East* (Printed by J.J. Little and Ives Co., New York) 1920, p. 126.

[3] Irving National Bank, p. 127.

Perhaps the most important precepts in marine insurance are found in Articles 17 and 18 of the British Marine Insurance Act, 1906, which only reinforced the long recognized legal necessity for the "utmost good faith in contracts for marine insurance" and imposed a duty "to disclose every material circumstance known, or presumed to be known, by the assured." If either of these requirements is breached, the insurer may avoid the contract.[4]

With all this in mind, we can now consider why many Norwegian shipowners chose not to carry marine insurance. First of all, having purchased 10- or 12-year old vessels, they could expect insurers to quote a higher premium than would be charged for new or recently constructed hulls.

While ship surveyors for the various classification agencies made regular inspections of all vessels as to seaworthiness, those examinations were normally conducted in port, usually at dry dock, where no stress comparable to a stormy sea was being exerted on the vessel. However careful a Lloyd's agent or inspector for other classification organizations might be, a potential defect could be overlooked under such conditions.

Presumably a shipowner investigating a bargain purchase of an older vessel would interview officers or crew members who had previously served on it, some of whom might volunteer (with a little liquid or some other tangible encouragement) important facts about the vessel or its sailing qualities. This information would have to be disclosed to an insuring agency.

There is, for instance, an oft-repeated story from years ago of a discharged seaman who swore to his drinking companions that the deck planks on the vessel he had just left were so loose that one of his boot heels had been caught in a vise-like grip between the butt ends of two planks during a heavy sea. He was freed several moments later when the vessel rode up a large swell and the planks changed position again. The inevitable result of such a conversation, if it were to come to the attention of a marine insurance writer, would have been a sizeable increase in premiums or in some cases coverage might have been arbitrarily declined.

To learn how insurance claims for maritime losses were handled in the *Struan*'s time, I have studied the 80-odd-page 1886 document entitled "Statement of General Average and Partial Loss in the Case

---

4 Leslie J. Buglass, *Marine Insurance Claims, American Law and Practice* (Cornell Maritime Press), p. 47.

of the Bark *Guiana*."[5] In the first pages is a description in handwritten script, of the voyage of the *Guiana* under Captain Eben C. Byrns from Iloilo on the island of Panay, Philippines, with a cargo of sugar, bound for New York. In his "protest," (a notification that damage had been suffered and insurance claimed, Byrns included excerpts from the official ship's log, to enable the average adjuster handling the case (and any other reader) to understand exactly what transpired.

Discussion of the *Guiana's* record of storm damage and related insurance negotiations and eventual settlement seems appropriate in view of the similarity of conditions under which the *Struan* sailed in the same year.

Though five years newer and about two hundred tons lighter in registered tonnage than the *Struan*, the *Guiana* was built at Tusket, Nova Scotia, just a few miles across the Bay of Fundy from where the *Struan* had been launched, and was considered an excellent example of superior contemporary wooden ship construction in its time. Another close tie to the *Struan* relates to the fact that both vessels, after a 12- or 14-year service with their respective Canadian or British owners, were ultimately sold to Norwegian interests, though in separate transactions and to different operators.

Additionally, the *Guiana* was commanded on its last voyage (it disappeared after leaving Mobile, Alabama for Bahia Blanca, Argentina on March 19, 1907) by Svend Hoie, who had been first mate on the *Struan* when it was lost in 1890.

Loaded with 31,200 piculs[6] of sugar and towed out into the bay

---

[5] This document was obtained from the archives of the Yarmouth County Historical Society, 22 Collins Street, Yarmouth, Nova Scotia, Canada B5A 3C8.

[6] A picul (equal to 100 catties) is a measurement of weight believed to have originated in Malaysia, used in China and other countries of the far east. At the time of the *Guiana's* voyage, the picul was falling out of favor with many traders because of the unwillingness or inability of users of the measuring system in different localities to agree on a commonly acceptable equivalent in standard pounds.

In the General Average statement prepared by Walker and Hughes after the *Guiana* suffered severe storm damage in 1886, the adjusters arrived at the cargo weight originally shipped by converting to pounds at *138-1/2 pounds per picul*, though through transposition of digits in their computation they made an arithmetic mistake, thereby reducing the weight of sugar actually transported (and affecting the actual freight due for that service) by 9,000 pounds.

Some dictionaries assign a weight to the picul of 133 1/2 pounds, while the Funk and Wagnall's *New Standard Encyclopedia* as late as 1934 ascribed to it a weight of 139.46 pounds.

The U.S. Census of the Philippine Islands, 1903, Volume IV, page 447 states simply that "the accepted weight of the picul in the Philippines varies from 86 pounds (39 kilograms) at Sorsogon, Luzon to 152 pounds (slightly over 69 kilograms) in Leyte."

by a steam tug on October 7, 1885 the *Guiana* was hardly under way before five of the crew fell ill with malaria, and four more became sick in short order.

Each day's entry tediously emphasizes that "pumps, sidelights and lookouts were carefully attended to." The vessel's progress is noted with almost daily recordings of latitude by observation and longitude by dead reckoning, and every recognizable headland and promontory is noted as to compass bearing and estimated distance to each landmark. On October 28, one of the able seamen died suddenly and was buried at sea.

On January 13, 1886 the *Guiana* stood in for Port Thornton, Ascension Island, and ran up signal flags asking for medical assistance. The answer from shore was "send a boat," and the master himself went ashore, where the doctor declined to board the *Guiana* but sent medicine for the ill on board. On February 4 the bark put into Carlisle Bay, Barbados to land and hospitalize two sick sailors and to obtain fresh water, provisions, and four replacement crewmen.

On February 16 the *Guiana* spoke the bark *Romanoff* of Yarmouth, Nova Scotia.[7]

On February 18 and 19 it "blew a whole gale" and "much water came over the deck." A terrific squall blew the jib and the mizzen staysail completely away. A very heavy sea on the 20th washed over the vessel and filled the deckhouse and cabins with water. The ship labored heavily, causing much leakage.

On February 19th, 1886 Captain Byrns recorded the *Guiana's* position as lat. 37° 36' N., long. 69° 36' W, only about five hundred miles off Delaware Bay on the U.S. east coast, and about six hundred or seven hundred nautical miles from his destination, New York City. Faced with the reality of a badly leaking vessel and a shortage of healthy seamen to work the sails for tacking into the prevailing heavy offshore winds while at the same time constantly manning the pumps,

---

[7] "Speaking" a vessel did not necessarily mean that of the crews actually communicated with each other by megaphone. It was too dangerous for heavy vessels at the whim of the wind to approach each other closely enough for conversation. A vessel was "spoken" when it passed near enough to the observing ship for the national flag and four-letter (four flag) identification signal to be recognized through marine glasses or a telescope.

The *Romanoff* was a 1049 net-registered-ton bark built at Shelburne, Nova Scotia in 1876 and registered at Yarmouth, the nearest port of Registry. It is possible that Captain Byrns of the *Guiana* recognized the *Romanoff* and was acquainted with its captain. Yarmouth is only a few miles from Tusket, Nova Scotia, where the *Guiana* was built.

Captain Byrns found himself forced to run 1,200 miles south to St. Thomas for repairs, in effect retracing a considerable distance of the voyage. The wisdom of that decision became apparent over the next few days as six more members of the crew were incapacitated by storm-related injuries.

With the decks continuously awash, the men were sometimes compelled to leave the pumps. The ship was making much more water than usual, and the gales continued until the 23rd, when one of the pumps broke down. On the 24th an exceedingly heavy sea washed over the vessel, disabling the carpenter and two seamen.

After a consultation with the other officers, and recognizing that the pump could not be properly repaired without a new casting, Captain Byrns decided to make for Bermuda, the nearest port of refuge where repairs could be effected.

The gale still continued on the 25th and 26th and a squall blew the topmast staysail out of is roping. Three more men were injured by the heavy seas.

On March 1, Byrns was able to make observations to plot the bark's actual position, and found that the storms had carried the vessel too far east of Bermuda to reach there with the existing strong westerly winds. The captain changed course for St. Thomas (now Charlotte Amalie) on the island of St. Thomas, West Indies.

In the next two weeks, even though the weather improved, the ship continued to leak at the rate of about 12 inches per hour, making a heavy burden for the remaining operable pumps until someone on board was able to jury-rig a temporary wooden valve to restore the pump with the broken casting to at least a part of its normal capacity.

On March 18 the *Guiana* finally limped into the anchorage outside St. Thomas where the vessel was put in charge of a pilot for entrance to the harbor itself.

Five days after arrival in St. Thomas, following an extended exchange of cables with the owners, conferences with the owners' agents, and approval of the underwriters, the vessel was surveyed by a master shipwright and an experienced master mariner. They found a myriad of problems, including leaking seams at a rate of two inches per hour at anchor in a protected harbor, a broken tiller, damaged rudder head, some other less serious problems, and recommended unloading the cargo for a full and complete survey.

Ten days later the British consul and a local merchant, after inspecting the vessel, suggested unloading the sugar cargo into other seaworthy ships to avoid inevitable losses from spillage if the cargo were first put into a warehouse and then trans-shipped at a later date.[8]

On April 21 yet another survey team reported that two offers for repair of the *Guiana* at St. Thomas had been received, for $13,502 and $12,290, either of which when added to consular fees, supervisory charges, contingencies, painting, etc., would total at least $15,000. It also reported that the captain and crew were willing to take the now unloaded vessel *in ballast* to Saint John or Halifax for repairs "where the cost would undoubtedly be substantially less" and that it was prudent and safe for them to do so, since a wind pump had been purchased and installed.

Captain Byrns brought the *Guiana* into Saint John less than three weeks after leaving St. Thomas. Another survey party, this time composed of members of the Board of Port Wardens, inspected the bark in "Quinn's Blocks"[9] and recommended stripping the metal hull covering to properly ascertain the full extent of damage. Two other specialists in assessing the value of vessels, determined the *Guiana* in her damaged condition, to be worth $18,500.

Yet another inspection team on May 27, after removal of the metalling, found seams and butts open, all treenails loose, several iron knee bolts loose in the lower hold and one anchor stock broken. They ordered that "all new rigging and sails be installed and that all damages be made good."

Treenails (pronounced tre'nals or trunn'als by shipbuilders and the maritime fraternity) were cylindrical wooden pins, similar to large dowels in present-day terminology, used to secure ship planking to the structural framework and in various other places aboard wooden vessels. In situations where treenails were driven into a "blind hole" (into but not through a connecting timber) small hardwood wedges were started in finely-cut saw kerfs in the end of the treenail to be inserted, so that the wedge would expand and confine the treenail as it was driven to the bottom of the bored hole. A barrel of 3,500

---

[8] This advice proved to be excellent. Substantial losses of sugar from torn sacks, etc., were realized in transferring the cargo directly from the *Guiana* to the two available vessels engaged to carry the cargo to New York from St. Thomas—the bark *Kreon* and the steamer *Benevista*.

A second handling would not only have cost extra labor, but would have resulted in further losses.

[9] Because of the extreme tidal variations at Saint John and other Bay of Fundy ports, operation of a float-in, float-out substitute for a dry dock was a practical solution for the inspection of vessels. If the vessel remained in the blocks for more than the interval between tides, block-tenders or tidesmen were employed to insure that the hull was properly supported in its blocks as the tide went out.

wedges was ordered for repair work on the *Guiana* at Saint John, but 3,000 were returned unused for credit.

In cases where a number of shorter pieces were joined together to form a continuous length piece of uniform dimension, as in preparation of a vessel's keelson, bolts would be used to hold the various sections in close proximity to each other, but treenails would also be used. These lightweight connections would not only strengthen the unit and provide rigidity, but would also serve to transfer shear forces as the action of the sea worked the vessel and continuously strained its joints.

In most cases, replacement of a loose treenail required that the old fitting be drilled out with a bit slightly larger than the original, in order that a new oversized treenail, made of dry hackmatack, locust or other suitable straight-grained hardwood, could be driven in with a mallet or heavy hammer. When the job was completed, the entire area was flooded or hosed down so that the dry treenails would absorb moisture and expand in their holes to make a very tight fit. Rebuilding the *Guiana* at Saint John in 1886 required the use of more than 3,000 treenails of various lengths and diameters.

Only three weeks later, the final survey found all recommended work completed, and the vessel was refloated and certified to be in seaworthy condition.

It wasn't until August 4th that the adjusters, Walker and Hughes, finished their paperwork concerning the *Guiana's* storm damage. Every expense incurred either at St. Thomas or in Saint John, however minute, had to be allocated or apportioned to its proper account in order that a fair settlement could be achieved. A 15¢ expenditure in St. Thomas for needles to sew up torn sugar sacks was as carefully noted as the $1.25 per night paid the watchman to guard the vessel in dry dock in Saint John, or the $1,500 purchase of new yellow metal to be installed after the hull repairs were completed. Some 25 double size pages were required for the final report.

Because the owners had insured the vessel for only one-half of its estimated value at the beginning of the voyage, their recovery amounted to only slightly over $2,800.

For some reason , the vessel's owners did insure themselves for the full amount of the expected freight, some $13,000 spread amongst a number of policies. Partly as a result of a General Average charge of $1,392 against freight, and because the *Guiana* had to off-load the cargo onto two other vessels at St. Thomas for delivery to the original destination and was therefore unable to earn the full amount of the originally agreed transportation fees, a total insurance recovery of only $1,400 was realized for vessel repair.

In this connection it is interesting to note that the firm which owned a major share of the *Guiana* was also in the business of selling various types of insurance, including marine policies. It must then be assumed then that they knew what they were doing, and that the actions taken in the matter of insurance on their vessel followed prudent practice at the time.

The *Guiana* was only 3 1/2 years old when it began its voyage from Iloilo. Built at a cost of about $45 per registered ton in 1882, it would have been depreciated by the owners, at about 15% annually, to the $30,000 pre-voyage figure put on it by the average adjusters at Saint John in the summer of 1886. Had that voyage from Iloilo been successful and without difficulty, the $12,500 freight on the sugar cargo carried would have returned the owners somewhere between $6,000 or $7,000 after voyage expenses were paid, reflecting an income of almost one quarter of the value of the vessel in less than half a year's sailing.

There was one distinct advantage that accrued to the owners, however. After the substantial repairs made to the *Guiana* at Saint John, they had a vessel in almost new condition—new sails, new rigging, retreenailing at all joints, recaulking and re-metalling. Barring an accident, this would have extended the life of the bark in seaworthy condition for another six or seven years.[10] The entire rebuilding project was completed in less than three weeks, at a total cost of under one-half the lower of two bids for the same work at St. Thomas.

What might have been the difference in costs to the owners if there had been no insurance on the *Guiana*?

From the $4,200 total actually paid toward the vessel's repair by the insuring companies, $1,000 in fees was deducted by the adjusters, Walker and Hughes for preparation of the "Statement of General Average," for commissions charged for collecting and disbursing funds, and for interest on the money calculated from the date the statement was prepared until the "probable date of collection"—some six months later.

An estimated $430 was paid as compensation to agents acting on behalf of the owners by telegraphic authority at St. Thomas, for arranging and supervising trans-shipment of the sugar cargo and dispatching the *Guiana* to Saint John.

There were also a number of special charges for negotiations with

---

[10] The *Guiana* did, in fact, continue in service for another 21 years until lost at sea on a voyage from Mobile to South America in 1907.

owners of the cargo (which perhaps under a no-insurance situation might just as well have been carried on by the *Guiana* owners), extra surveys and inspections (all involving delays) made to conform to the requirements of the underwriters and the various contracts of insurance involving some seven different companies, totalling perhaps $550.

The amount of the premiums expended for coverage under the multiple policies is hard to substantiate, but a conservative figure would be $1,480, based on our understanding of applicable rates in effect during the year in question.

Subtracting all the extra costs that would *not* have been incurred had there been no insurance in force, the bark owners' would have netted the relatively insignificant sum even at that time, of $740.

So, it is not difficult to understand the rationale of many Norwegian shipowners of the same period in dispatching their vessels without insurance coverage. A typical Norwegian shipowner in 1886 could have used that entire episode of the *Guiana* as an argument for "going bare."

Byrns' main problem, the "skipsreder" would argue, was a shortage of personnel to manually operate the ship's pumps from deck stations during stormy weather over a long period of time. The *Guiana's* log makes it clear, he would insist, that several crew members were injured by heavy seas while engaged in those duties. If the *Guiana* had been equipped with a wind-powered pumping system favored by many Norwegian mariners, crew members would have been available to help handle the ship when all hands were needed for sailing close-hauled into an offshore wind the last seven hundred miles of the voyage, instead of retreating some 1,200 miles to a port of refuge.

Our mythical Norwegian shipmaster would be pleased to learn that a wind pump was indeed purchased in St. Thomas and installed in the *Guiana* before Captain Byrns left for Saint John in ballast; but he would not have understood why it was removed during repairs and sold at auction in Saint John before the vessel was returned to service.

Norwegian "skipsreders" were aware long before the time of the *Struan* and the *Guiana,* that their investments in ships consistently returned them greater profits on invested capital than did any other industry. With older vessels and lower labor rates, owners could afford to take an occasional risk and still realize as good a return as could be expected from any other pursuit.

Norwegian vessel operators, however, were not the only ones who sent their fleets to sea without insurance protection. The *Halifax Chronicle* in the early summer of 1877 called attention to a shipowner named Gideon Palmer of Dorchester, New Brunswick, who for more than twenty years, sent to sea at least six vessels of up to 1,000 net register tons, and never lost a single one. His practice was to maintain the size of his fleet at about six by frequently adding new vessels and selling the oldest. "This remarkable fact," the newspaper editorialized, "unparalleled perhaps in maritime annals, seems to show that either Providence or Mr. Palmer's captains exercise peculiar watchfulness over his interests."[11]

[11] Reprinted by the *Saint John Globe*, August 14, 1877

Chapter XVI

A Change of Flags

In the late summer of 1889, Andrew Gibson of Liverpool was approached by representatives of a group of shipowners from Haugesund, Norway who were interested in purchasing the *Struan*. Realizing that the ship was over 12 years old and that the Norwegian syndicate seemed prepared to offer a fair price (estimated at about 65% to 70% of the original cost when built in 1877,) Gibson agreed to the sale and directed a letter to the keeper of the registry lists authorizing transfer from British to Norwegian registry. Accordingly, the *Struan*'s British registry was closed on November 12, 1889 and registration under the Norwegian flag became effective November 22, 1889,

A newspaper article in the *Haugesunderen* in September 1889 announced the purchase of the *Struan* to the approximate 3,000 citizens of Haugesund:

> VESSEL PURCHASED. The Struan, about 1470 registered tons, has been purchased by joint owners Erich Lindoe, E.H. Kongshavn, H.O. Sundfor, A. Lothe, Christian Christiansen, A. Anda and others. The ship is rated first class by Lloyds of London. Purchase price is reported to be 7700 English pounds. The vessel will be commanded by T.H. Skogland, the earlier skipper of the schooner-brig *Sigurd*. The ship's coordinator is shipowner Mr. Erich Lindoe.

An old ledger sheet obtained from the grandson of E. H. Kongshavn reflects the purchase of a five percent (1/20th) share of the *Struan* in the Fall of 1889 for the sum of 7,200 krone. At the rate of exchange then in effect, that amount would translate to a total vessel value of about 7,921 British pounds.[1] Against that outlay, the page shows only a single 500 krone income item from the *Struan*, apparently a distribution on October 14, 1890 of the freight earned for transportation of the coal cargo from Birkenhead to Iquique, less perhaps the pro rata cost of storm damage repairs and installation of the wind pump at Plymouth, England.

The proportionate interest of the other Norwegian owners in the *Struan* is not known, but it can be assumed that Erich Lindoe (the youngest of the group), was designated managing owner because of his having a major, or a controlling, share in the ship. Such an investment, of course, meant he would suffer the greatest loss should the vessel be lost.

Lindoe owned interests in several other vessels, and received so many telegraphic messages that he recorded the cable address *Eric, Haugesund* in Lloyds' Register of Shipping each year to facilitate prompt and economical receipt of his telegrams.

The difference between the price given in the newspaper and the actual purchase price in Liverpool is explained by the fact that the ship was put into drydock immediately after the change in ownership to have the hull recovered with new metal. Deck boy Anton Madsen in his later recollection refers to the re-metalling as being done with copper, but Lloyds' records reflect the use of "yellow metal," at that time a trade term on the continent for what is now known as "Muntz Metal," an alloy of 60% copper and 40% zinc, invented in 1832 by George F. Muntz. The advantage of Muntz metal over ordinary copper for covering the hulls of ships was its greater hardness. It resisted scraping, scratching and abrasion better than pure copper, a much softer material, but may not have been as effective in delivering the anti-fouling protection that shipowners expected. Chemical engineers and metallurgists have long known that pure copper, or alloys containing 85% or more copper provide the greatest resistance to fouling and are usually preferred when superior anti-fouling protection is particularly

---

[1] We are indebted to Norges Bank (Bank of Norway) which has made available to us through John M. Hauge III (grandson of the third mate of the *Struan* at the time of its loss), a copy of the published rates in effect for foreign currency transactions on October 31, 1889. Included is the information that:

Nkr 18.18 = 1 British pound.

In answer to an inquiry by the captain of the *Tamar E. Marshall*, the British Consulate General's office in New York wrote on 15 March, 1892:

"I hereby certify that the average rate of exchange...[on the date of inquiry] of bills on England, at three days sight, was US $4.88 to the pound." Signed Horace D. Nugent, vice consul.

Since Nkr 18.18 = US $4.88, we can ascertain the then-existing values of the krone to the U.S. dollar, and vice versa:

US $1.00 = Nkr 3.725; and Nkr 1 = US $.2684.

The *Struan* under sail in 1890. Note the extremely fine detail, including two sailors at work high on the foremast furling a sail, and the capstan on the forecastle (fo'c's'le) head, just abreast of the forward name board.

Under agreement with the Swedish government, the Norwegian flag in use at that time (until 1898) included in the upper staff quadrant additional insignia then commonly referred to as "sildesalat" (sliced pickled herring, beets, onions, etc.). Embassies and consulates were jointly operated for serving the citizens of both countries. Some artists not familiar with this detail flagged their Norwegian ship paintings covering the period 1844 to 1898 with the present day Norwegian ensign, which did not come into use until after 1898.

No identification of the artist is given. The painting may have been done by a "pier-head" artist at Liverpool, Birkenhead or Plymouth, all British ports visited by the *Struan* during its registration under the Norwegian flag, or could have been commissioned by one of the joint owners either before or after the vessel's loss in December 1890, and executed by a Norwegian painter.

— *From the collection of shipowner E. H. Kongshavn, Haugesund, Norway (home port of the Struan at the time of its loss). Used by permission.*

A reproduction of this painting also appears on the book jacket.

important. The function of the metal hull-covering is the slow release of copper ions that poison barnacles and other organisms as the vessel moves through the water. Such action protects the copper-covered hull itself, but does not shield adjacent non-copper surfaces.

In an effort to reduce the problems of bottom-fouling and sea-worm damage to the hulls of British vessels of war, the Admiralty in 1761 began experimenting with copper sheathing by covering the hull of a 33-gun frigate. Four years later, they similarly treated another vessel. By 1777, 14 more craft joined the test program, but the entire fleet did not get copper sheathing until the mid 1780s. Because of the cost, owners of merchant vessels were slow to follow the practice, though copper sheathing of all wooden hulls was well established by the turn of the century.[2]

Because some iron bolts and fastenings had been used in the original construction of the *Struan*, it was necessary to first apply a heavy coat of felt over the hull planking to avoid the possibility of galvanic or chemical action between the two metals.

A study of old records of sailing ships reveals that vessels were normally re-metalled every two or three years; any longer period between replacements resulted in loss of speed and maneuverability. In tropical or warmer ocean environments, hordes of marine organisms, mostly barnacles, would attach themselves to the exposed patches of wooden hull where the metal had been torn away, whereas they could not easily fasten themselves to the smooth metal itself.

Captain Tollak Johan Hansen Skogland, 32, was appointed by the owners to command the *Struan*. He had previously been skipper of the Norwegian schooner-brig *Sigurd*, and brought with him the long-time first mate of that vessel, Svend Hoie, to fill the same position aboard the *Struan*. Instead of hiring a crew in Liverpool, Captain Skogland had the entire ship's company sent from Haugesund as a unit, many crew members having been selected for their particular talents or abilities. In those days before motor-driven or engine-powered small boats, every skipper had to have a sufficient number of capable small boat handlers among his crew. Unlike present-day U.S. vessel-manning requirements, which dictate the presence of a certain number of specially rated lifeboatmen in each vessel's crew, maritime regulations a hundred years ago left such matters largely to the captains themselves.

---

[2] J. Richard Steffy, *Wooden Shipbuilding and the Interpretation of Shipwrecks* (College Station, Texas: Texas A & M University Press, 1994) p. 175.

Every time a vessel dropped anchor in a port of call, the captain had to call out a boat crew to take him to the local port authority office to deposit the ship's papers and make arrangements for unloading and/or loading cargo, etc. Sometimes these trips had to be made in relatively bad weather, and boat crews were necessarily required to be competent under all conditions. Accordingly, Captain Skogland and his first officer were careful to include in the crew a number of men they knew to be expert in the handling and care of small boats. Many had experience in the Haugesund fishing fleet, and most had begun their acquaintance with boats as children. The wisdom of including an adequate number of competent boatmen in the *Struan* crew would be demonstrated more than once during the voyage.

Skogland himself came from a farm at Skare a few miles south of Haugesund. First mate Hoie was a native of Avaldsnes on the island of Karmoy, located close to the mainland south of Haugesund. Other crew members came mostly from the Haugesund area, with a few living along the fjords not far from the city. We have not been successful in determining the birthplace or previous experience of Thomas Jorgensen, engaged as second mate.

While the captain busied himself with inspection and supervision of the work being done to his vessel during the re-metalling, and making final arrangements for cargo loading, Hoie went about shipping the necessary supplies for the trip. Food, rope, spare sails and spars, paint, drinking water and a hundred other items had to be obtained and put aboard. Navigational charts, if left aboard by the previous owners, had to be checked to make certain that they were up-to-date and that the collection was complete, since there was no way of knowing what part of the world the *Struan* would be visiting or in what waters the ship would be sailing after unloading the first cargo.

Since the invention and installation of telegraphic communication facilities at most ports during the twenty years preceding the *Struan*'s transfer to Norwegian registry, it was customary for vessels to "call for orders" by telegraph from wherever a cargo was discharged or at the nearest port having telegraphic capabilities.

Although there are, to our knowledge, no existing documents to verify the weight of coal loaded at Birkenhead (across the Mersey river from Liverpool) for the first voyage under the Norwegian flag, deck boy Anton Th. Madsen's later recollection of the trip relates that a "full load" of coal, 2,650 gross tons, was taken aboard. While such a load reflects a slightly higher cargo weight to registered tonnage ratio than even the *Canada* cargo discussed as an example of a very heavy load on page 77, it seems well within the realm of possibility, and we have no basis for questioning the figure Madsen mentions.

At about 87 pounds per cubic foot, such a coal cargo would theoretically have occupied about 68,230 cubic feet of the 147,300 cubic feet—slightly less than one-half—of hold space available for stowage of cargo aboard the *Struan*.

It should be kept in mind that the cargo was put aboard at a river port where the salinity of the water became a factor. The Plimsoll mark or draft disc required by Parliamentary action for all British vessels some 12 years previously, and by most other maritime nations' governments soon after, involved a load limit based on salt water draft. Accordingly, if the water at the loading pier was fresh or almost fresh, the vessel could have been legally loaded to a slightly greater depth than the Plimsoll mark indicated, since the hull would have "come up" perceptibly upon entering salt water.[3]

While managing owner Erich Lindoe had arranged from Haugesund the first cargo for the newly Norwegian-registered *Struan*, it was Skogland's responsibility to finalize cargo arrangements and to make certain that his officers supervised the actual loading to insure that the load was distributed in the hold in such a manner that the vessel was kept in trim, not too heavy fore or aft, and with no list.

The first knowledge that he would be loading an entire coal cargo for Iquique, Chile must have given Skogland something to think about. A long trip with a shipload of coal, particularly through the tropics, was not like a similar voyage with a noncombustible cargo. (Coal's propensity for ignition by spontaneous combustion has been discussed in the chapter on cargo.)

Upon completion of loading, the *Struan* departed the Mersey river near the end of November 1889, and according to deck boy Madsen's later chronicle, headed almost instantly into bad weather and rough seas. Before long the ship sprang a leak, and continuous work at the pumps was required. Without the vessel's logbook, it cannot be determined exactly how far south the ship had progressed on its voyage by the time it became necessary to turn back and seek the nearest port to

---

[3] One cubic foot of fresh water weighs 62.5 pounds (1,000 ounces). One cubic foot of salt water weighs 64 pounds (1,025 ounces). Salinity testers, together with formulas and tables for determining the exact amount of draft overload that can be made in water of a certain salt-fresh mixture, are available to shipmasters today to facilitate loading under all circumstances. Whether or not such equipment was available one hundred years ago, or whether such a computation then was a "seat of the pants" guess is not known. It is presumed that shipmasters routinely made local inquiries regarding the amount of salt in the water at their port of loading.

effect repairs. It is known that sometime after taking that action the *Struan*, on January 11, 1890, in latitude 48° N and longitude 8° W[4] came upon the Dutch vessel *Louise* in sinking condition, with all its boats smashed by the storms. In a rough sea with a gale blowing, the *Struan*'s boatmen were able "with some difficulty" to rescue all hands from the Dutch vessel almost as it sank under them.

Madsen identified the rescued crewmen as Russian citizens, though he could not, as a first voyage apprentice, have been expected to differentiate between the national flags of The Netherlands and Russia. Both included the same red, white and blue horizontal stripes, though in different configurations. The Netherlands flag had equal-width horizontal stripes of red, white and blue, with the red stripe at the top. The Russian flag in use at the time had a red stripe at the bottom, a blue stripe in the center, and a white stripe at the top.

It is easy to understand how confusion about identity of the *Louise*'s flag could have arisen, particularly if the captain of that vessel had hoisted his ensign upside down as a signal of distress. If the flag were partially wrapped around its halyard, its identity would be difficult even for an expert, much less by boatmen concentrating on saving the lives of the crew of the sinking vessel.

The *Encyclopaedia Britannica* in its 11th edition (1910), volume 10, page 463, cites a practice "known to be in use as far back as the reign of James I" (1603-1625):

> When a ship has to make a signal of distress it will hoist its national flag upside down. To accentuate the imminence of danger, the flag can be made into a 'weft' by knotting it in the middle.

If such a procedure was known to and utilized by the *Louise*'s captain, his flag would have appeared to *Struan* crew members as a jumble of red, white and blue intermingled colors, almost completely indistinguishable to any observer.

It may well be that the Russian identification was made for him by others in the *Struan* crew, upon whose experience and knowledge he depended. His account of the incident recalls that the "eleven rescued mariners"[5] were "of great help with the *Struan*'s pumps."

It must have been somewhat disconcerting for the sailors from the *Louise* after laboring desperately for three days and two nights in a vain attempt to keep their own vessel afloat, to find themselves aboard

---

[4] About 150 nautical miles off the coast of Brest, France.

a much larger rescue craft which itself was taking on water at a rate probably exceeding what had been encountered aboard the *Louise* especially since they would spend more than two full weeks after the loss of one vessel taking their turns at the *Struan*'s pumps before that ship reached a port of refuge.

In any event, the *Struan* arrived safely at Plymouth, England on January 25, 1890, and the *Western Daily Mercury* two days later announced the names and ratings of the rescued Dutch mariners. Information regarding the incident also appeared in the January 25 edition of *Shippers' Gazette and Lloyds' List* (published in London).

It is very likely that *Louise* (not *Louisa*) was the correct name of the vessel from which the *Struan* boat crew rescued eight distressed mariners in early 1890. The Royal Netherlands Embassy in Washington, D.C. has made available information confirming that there was a relatively small Dutch bark (883 cubic meters, call letters P J K B) named *Louise* listed in the 1890 edition of *Lijst van Nederlandsche Oorlogs-en Koopvaardijschepen* (*List of Netherlands War and Merchant Ships, 1890*).

From official records, the *Struan*'s measurement in cubic meters was 4,169. The *Louise* would have contained approximately 1/5th the cargo space of the *Struan*, being perhaps slightly over one hundred feet in length.

Inquiries to the Schiedam archives, the Prins Hendrik Maritime Museum in Rotterdam and the Rijksmuseum Nederlands Scheepvaart Museum in Amsterdam revealed that (1) no mention was made in any story in Dutch newspapers concerning the *Louise* near the time of the incident; (2) names recorded in the Plymouth, England news story identifying crew members "do not sound Schiedams," and that the crew and the captain "were in all likelihood mustered elsewhere"; and (3) the limited records surviving from the reported owners and operators of the vessel (A. Prins & Co., shipowners and shipbrokers, 1793-1920), make no mention of the *Louise*, its loss, or the rescue of the crew.

In connection with these investigations, it was learned that Dutch shipping law did not require masters of vessels to be licensed until 1902, and that "anyone could be a ship's captain" prior to that date.

One of the original *Struan* crew members was replaced at

---

[5] Captain Skogland reported saving the entire crew of the *Louise* and brought eight men to Plymouth. No mention was made in newspaper accounts at the time of the rescue about any additional crew members who had not been rescued. We must assume Madsen's "11" was an inflated figure.

WESTERN DAILY MERCURY

Plymouth, Devon, January 27, 1890

The crew of the barque *Louisa*, of Schiedam, Holland, Captain G. Frew, bound from Swansea to Hayti, West Indies, with a cargo of patent fuel, were landed at Plymouth yesterday by the Norwegian ship *Struan*, 1,473 tons register, Captain Skogland, bound from Liverpool to Iquique with a cargo of coals, which put back to effect repairs. The *Louisa* left Swansea on December 27 with a crew of eight all told. She experienced terrific weather, and after being about fifteen days at sea the vessel began to make water rapidly. The crew were ordered by the captain to work the pumps, and they continued pumping for three days and two nights. The water gradually increased in spite of the efforts of the crew. Signals of distress were hoisted and observed by the *Struan*, which hove–to and took off the captain and crew of the *Louisa*, who were in an exhausted condition. The *Louisa* was abandoned on January 11 in lat. 48 N. long. 8 W. in a sinking condition, having six feet of water in her hold, with all her boats smashed. It is believed she foundered shortly after. The crew were treated with every kindness on board the *Struan* by the crew of that vessel. The men when landed at Plymouth were taken to the sailors' Home and received the prompt attention of the matron, Mrs. Phillips. The names of the men are G. Frew, captain; G. Blaase, mate; G. Janson, second mate; E. Lappen and O. Fiederman, able seamen; E. Rassick and J. Wondoivick, ordinary seamen; and E. Born, cook.

Reconstruction of an 1890 news article with corrections made to eliminate originally published errors.

Plymouth. Captain Skogland had reported on his arrival there that one officer and three other hands had been injured in storm-related accidents during the voyage from Birkenhead. Whether the replacement of Pedar Olai Anderson Hauge[6] by Ingvald Ommundsen Odland was related to injuries at sea cannot be established with available records, but P.O.A. Hauge deserted at Plymouth, and Odland was hired in his place the same day.[7]

[6] This seaman, who lost his life aboard the bark *Assistant* of Kingston, Jamaica on a voyage from Mobile to Jamaica in December 1893, was, so far as can be determined, not related to the John Mandius Hauge who joined the *Struan* at Port Townsend in October 1890.

[7] How Odland happened to be available for employment at Plymouth on the very day that Hauge deserted is something of a mystery. Odland's official seaman's records indicate that he signed off the vessel on which he had previously served at Stavanger, Norway on December 5, 1889. The next entry in his record relates to service on the *Struan*, begun at Plymouth 11 February 1890. No explanation is evident of how he reached Plymouth, since no employment is recorded between 5 December 1889 and 11 February 1890. He may have come to Plymouth as a passenger on commercial transportation upon having been summoned by a ship's officer after the *Struan's* arrival in Plymouth on January 25th, 1890.

Chapter XVII

The Final Voyage

After completing storm damage repairs and departing from Plymouth on February 14, 1890, the *Struan* didn't meet another vessel at sea for almost two months. Sometime previous to April 15, she was spoken off Cape Horn but didn't arrive at Iquique until June 16.

The elapsed times for the various legs of that voyage warrant some comment. The approximate 60 days of sailing time to cover the 8,000 or so nautical miles from Plymouth to the Cape Horn area (averaging 133 miles per day, or about 5-1/2 knots) would have been reasonable at that time of year. But why it took two months (April 15 to June 16) to cover the approximately 2,400 mile trip from the Cape Horn vicinity to Iquique, however, cannot be explained without reference to the *Struan*'s logbook, which has not been located in spite of a long-term search. It is thought that Norwegian law at the time may have permitted captains to keep their logbooks after completing a voyage.

Recalling that sailing vessels had to deal with the strong northerly Humboldt current and prevailing winds from the south when navigating the west coast of South America, the *Struan* may have overshot its destination and been forced to make a wide counter-clockwise loop to approach a second time from the south. Lubbock relates the experience of the *Lindisfarne*,[1] a British sailing vessel which experienced the same difficulty there, resulting in an additional month's sailing time.

Knowing from the record that about two years earlier Captain John Scott had sailed the *Struan* (then under British registry) from Valparaiso to Pisagua, a passage almost half the distance from the Cape Horn area to Iquique along the same coast, in only nine days,

[1] Basil Lubbock, *The Nitrate Clippers* (Glasgow, Scotland: Brown, Son & Ferguson, Ltd., Nautical Publishers, 1932) p. 10.

suggests that some navigational misadventure may have befallen Captain Skogland's ship. If we consider three weeks an adequate time for that portion of the trip, and allow an extra week for good measure, there would still have been a full month available for a circular loop duplicating the *Lindisfarne's* unfortunate experience.

Deck boy Madsen's belated history of the voyage (in about 1935) does not mention any such occurrence, but it is easy to understand that crew members in that time were not always privy to information concerning the actual day-to-day location of the vessel on which they served. If he did at one time know of such an incident, Madsen may have forgotten about it in the 45 intervening years between his service aboard the *Struan* and his recollection of the voyage history; or he may have purposely refrained from mentioning it to avoid embarrassment to the living members of Captain Skogland's family.

The harbor at Iquique was what is known as an open roadstead, only partially protected by nearby Serrano Island. Developing from a tiny Peruvian fishing village in 1830, the city with its 5,000 inhabitants by 1856 had become the second largest port in Peru. It was blockaded by the Chilean fleet in 1879 and was included in the Peruvian province of Atacama annexed by Chile under terms of the Treaty of Ancon in 1883 when hostilities ceased. Though Bolivia by that settlement was denied an ocean port on its own territory, Chile contributed the equivalent of 25 million pounds sterling (some 120 million in U.S. dollars) toward construction of a railway from LaPaz, Bolivia to the Chilean port city of Arica.[2]

Lubbock's brief but vivid description of the Northern ports on the west coast of South America—that they were remembered by sailing ship seamen as "steep treeless hills simmering in heat, ankle deep in dust and swarming with flies"[3] is unequalled. Iquique did have one distinct advantage over the other ports, however—it was comparatively free from fog, though it averaged a 75 percent humidity throughout most of the year.

Iquique had been almost destroyed by an earthquake in 1866, was swept by earthquake-induced waves in 1868, and was subjected to earthquakes, fires and tidal waves in 1877.

Excellent photographs of sailing ships anchored off Iquique in a

---

[2] Earl Chapin May, *2,000 Miles Through Chile* (New York, The Century Co., 1924) p. 430.

[3] Lubbock, p. 1.

Hans Christian Baardsen, born in Haugesund June 18, 1872, deserted from the *Struan* at Iquique, Chile on August 14, 1890 while the ship was unloading a coal cargo. While working his way back to Norway, he had this photo taken at Boston in 1890 or 1891. Could the watch and chain visible in his vest pocket be the same one visible in the photograph opposite?
— *Photograph courtesy Helen Baardsen Conway*

Upon returning to Norway, Baardsen continued his seafaring career; passed his first mate's test on December 7, 1893 and received his captain's papers on December 4, 1897. This photograph, reconstructed from a group picture of his vessel's entire crew, dates from the time he commanded the small steamer *Haegholmen*, making several trips to England and Scotland from Haugesund in 1898-99. In later life, Baardsen lost the sight of one eye, probably as a result of repeated use of his sextant in bright sunlight, perhaps with the smoked glass protective eyepiece either missing or not properly in place. He passed away in May 1950.

— Photo courtesy of Helen Baardsen Conway

THE MORNING LEADER
Port Townsend, Washington

Friday, October 10, 1890

The British bark *Tamar E. Marshall* arrived yesterday from Yokohama. She will load wheat at Tacoma for the United Kingdom.

The Norwegian ship *Thor*, which is completing her load of lumber in Tacoma for London, will carry away a cargo that in some respects is the most remarkable ever sent from the Sound. It includes a greater number of long, heavy, sawed timbers than were ever placed in a ship's hold in this state. The last heavy timbers, loaded by means of a donkey engine, were each two feet square and from 107 to 110 feet long. There are yet about 450,000 feet of lumber to be placed on board. It is supposed it will take several months to unload her, and it has already taken sixty days to get the timber on board.

Tuesday, November 18, 1890

The Norwegian ship *Struan* is lying loaded at (Port) Discovery. She will clear from here today.

Tuesday, November 25, 1890

AT TACOMA: The British bark *Tamar E. Marshall*, 1,270 tons, Captain Parker, loading wheat for the United Kingdom. She is one of 56 vessels loading at local and Puget Sound ports this week, a big increase from the previous week.

Verbatim extracts from *The Morning Leader* 1890 news articles.

picket line formation with almost military precision exist in an earlier history of the area.[4] Vessels lay bow-to-the-sea perpendicular to the shore, about two hundred feet apart, each with two heavy anchors out ahead and one astern, to keep them in place (their captains hoped) against the possible ravages of giant waves resulting from undersea earthquakes offshore. There are recorded instances in which ships sometimes dragged their anchors under such attacks by nature, creating serious collision hazards with nearby vessels. In a few cases where anchors held, vessels were lifted so suddenly that anchor chains parted or capstans or anchor windlasses and even hawsepipes were ripped out of a ship, permitting filling the hull with seawater and subsequent loss of the vessel.

A study of enlarged copies of these photos indicate that vessels were moored bow-to-shore at the nearby port of Pisagua, where the feared fierce winds swept down without warning from the nearby mountains and presented an even greater threat than did earthquake-induced waves at Iquique.

Unloading was accomplished by transporting cargo to shore in small lighters, each crewed by two oarsmen or scullers. Similarly, anything to be loaded aboard ship had to be handled in the same manner, contributing to what some captains considered excessive handling time.

Within twenty years after the *Struan* unloaded its coal cargo from Birkenhead at Iquique, there were 14 coal mines being operated in Chile, employing some 12,000 miners producing substantial quantities of fair-grade coal suitable for use in steam boilers.[5]

Iquique was a port where fish and a variety of shellfish were plentiful and easily obtained, a welcome respite for the *Struan* crew from the ration of salt pork and pickled herring usually served on a sailing vessel in those days six months after departing on a lengthy voyage.

By consuming only the varieties of fish eaten by the local population, *Struan* crew members were spared the problem of determining the edibility of the species hooked by the method usually employed. In his memoirs of being shanghaied aboard the British bark *Musselcraig* at Astoria, Oregon and his subsequent nine-month involuntary service until rescued by a police launch at Sydney, N.S.W., Australia, W.J.R. Osborne told of the practice of placing a shiny silver

---

[4] Lubbock, opposite p. 50. A 3-leaf fold-out photograph, apparently taken from a nearby headland, with a quality and clarity of detail equivalent to the best of today's aerial photography.

[5] May, p. 362.

coin in the flesh of the freshly caught fish to be tested—in his case a bonita. If the coin remained clean and bright after three or four hours, it was believed the fish could be safely cooked and eaten. If it were even slightly discolored or tarnished, the fish would be considered poisonous and discarded.[6]

Even in prehistoric times the desert coast of Chile and Peru had supported primitive fisherfolk in scattered villages where tiny ribbons of fresh water fed by melting snow in the distant Andes made habitation possible. Ancient fish hooks filed down from mussel shells have been discovered in the area. Located on a relatively high desert plateau, the town now depends for its water supply on a 75-mile long pipeline from the oasis at Pica. In the *Struan*'s time, a barrel of clear fresh water was a commodity of high value in the area, and supplying a departing vessel with fresh water for an extended voyage became a challenge for shipmasters.

As coal unloading was almost completed and preparation for departure began, a second desertion from the *Struan* crew took place. On August 10, 1890, seaman Hans Christian Baardsen was recorded as a deserter when he did not appear for scheduled duty. Baardsen was known to have been interested in learning navigation, and it is possible that he may have joined some other vessel whose captain, facing a serious crew shortage because of multiple desertions from his own ship's company, was willing to promise personal instruction in navigation in order to get another competent seaman on his roster before leaving Iquique.

No further trace of Baardsen appears in Norwegian seamens' records until March 10, 1892, when he signed aboard the *Jokul* at Haugesund for a less than 30-day roundtrip voyage to England. No explanation of his activity during the 19-month period between his leaving the *Struan* at Iquique and his re-employment as a Norwegian seaman has been suggested, and it would be necessary to know the name of the vessels on which he served and the flags under which he sailed to search out that information. Baardsen qualified for a Norwegian mate's ticket on December 7, 1893, and passed his tests to be a Norwegian ship captain on December 4, 1897. In later years he opened a ship-chandlery shop on Haugesund's bustling waterfront. That business, although under different ownership, continues to serve the community under the Baardsen name.

---

6 W. J. R. Osborne, *Shanghaied at Astoria* (In *Cumtux*, Clatsop County Historical Society Quarterly, Vol. 13, No. 3: Summer, 1993) p. 6.

Captain Skogland, anxious to leave Iquique, arranged for ballast to be taken aboard his vessel, and finding no suitable replacement for Baardsen in the pool of available job seekers there, slipped away from the Iquique anchorage and sailed north for Port Townsend, Washington on August 14, 1890.

It cannot be determined, from available existing records, whether Skogland actually intended to take the *Struan* to Vancouver, British Columbia, Canada for loading there (as deck boy Madsen suggested in his later written recollection of the voyage) or whether the reference to Vancouver in Madsen's notes resulted from the captain's making a vague reference to Vancouver because it was more familiar to the crew.

In contrast to Madsen's comments, however, the October 10, 1890 edition of *Lloyds Weekly Shipping Index* (published in London) addressed the *Struan* as having left Iquique August 14 for *Victoria (Cal.)* [*sic*]. In retrospect, it appears likely the destination given should have been Victoria, British Columbia (B. Col.) and that errors in transcribing or transmitting the telegraphic information occurred. "Cal." would have resulted from "B" being omitted and the small "o" shown as "a." The state of California doesn't have a port named Victoria.

This confusion in the identification of the *Struan*'s destination supports our position in a previous chapter that information appearing in *Lloyds Weekly Shipping Index* was probably printed verbatim as received, without any effort being made to check reports for accuracy or to make corrections or adjustments before publication. In light of that conclusion, we have not accepted "index" reports as factual unless their information can be verified by other available sources.

Recalling that the *Struan* (then under British registry) had visited the Vancouver, B.C. area a few years earlier, and had loaded a timber cargo in the Moodyville-Port Moody district for delivery at Melbourne, Australia, it may be that Skogland's intended destination upon leaving Iquique was actually a British Columbia port. After entering the Strait of Juan de Fuca about September 29, 1890, a stop at Port Townsend would have taken the *Struan* only a very few miles out of its way toward a Canadian landfall; an investigation of the mill situation while there and telegraphic communication with the managing owner in Haugesund may have resulted in previous plans being altered.

The popular meeting place in 1890 Port Townsend for captains to meet with other members of the maritime fraternity for arranging cargoes, charters, dockage, towage and insurance matters, etc., was

Adrift at Sea.

The schooner Antelope arrived in the bay last Monday, and the captain reports that a day or two previous he sighted a full rigged ship adrift about 100 miles northwest of Coos bay. When he first discovered her, the main mast was broken off about 5 feet above the deck, and her fore and mizzen masts were standing, with top-sails set. As it was late in the evening, and a heavy sea running, the Antelope stood off, but next morning again fell in with her. This time the mizzen mast and light spars were down, and the whole mass was trailing astern. He made out her name to be the Struan, and says she is copper plated and looked like a new vessel. She was floating high out of water, indicating that her cargo was of lumber. No reason could be assigned for the crew deserting her.

The following dispatches appeared in the Oregonian, and probably refer to the same vessel.

San Francisco, Dec. 23 -- A Tacoma dispatch to the Merchants' Exchange says the captain of the ship Ericsson reports having passed on December 19, thirty miles off Cape St. George, an abandoned ship, with her main and mizzenmasts gone. The vessel was standing in shore when last seen.

Coos Bay News, Coos Bay, Oregon—Wednesday, December 31, 1890.

George Sterming's recently opened Belmont Restaurant and Saloon, located in the new block he had built the previous year and fronted with locally-produced cast iron pilasters to lend an air of Victorian elegance to his premises. Most major countries maintained consular representatives at Port Townsend, and it was in the joint Swedish-Norwegian consulate there on October 9, 1890 that Captain Skogland engaged as third mate on the *Struan* John Mandius Hauge, a 22-year old Norwegian seaman who was in the area, according to family recollections, after having visited a relative who had moved to a remote Alaskan settlement from Norway some years previously.

Deck boy Anton Th. Madsen's account of his voyage on the *Struan* refers to the run from Iquique to Port Townsend with enthusiasm not found in other parts of his narrative, perhaps due to a spell of good weather during the passage. "With a fair wind the ship ran fast, making up to 13 knots; we got an eye for the *Struan* being a good sailer," he recalled. "In forty days we arrived at Port Discovery."[7]

Although the *Struan* had reached the Port Townsend anchorage before the beginning of October, negotiations and arrangements for a timber cargo to Melbourne required considerable time. The first indication that loading had begun came in a notice in the marine section of the Tacoma Sunday Ledger, November 9 1890:

> At Port Discovery [a townsite on Discovery Bay just east of Port Townsend] the Norwegian ship *Struan* is loading lumber for Melbourne.

Loading continued until about November 18. On November 22, the *Struan* cleared customs, and departed for Australia on November 24 with 1,203,569 fbm (feet board measure) of rough lumber (mostly larger timbers) valued at $12,055.60; 57,800 laths valued at $144.40; and 52,320 pickets worth $523.20.

The British-flagged bark *Tamar E. Marshall*, built and owned in Digby, Nova Scotia, Canada, arrived in the Port Townsend area shortly after the *Struan* and was scheduled to load sacked wheat for Sunderland, England.

It is possible that during the three- or four-week period both vessels were at Port Townsend and Port Discovery, Captain Skogland of the *Struan* and Captain Parker of the *Tamar E. Marshall* may have

[7] Anton Th. Madsen, deck boy on board the *Struan* during the ship's final voyage in 1890, from his recollections of the trip written in Norwegian for the *Haugesunds Rederiforenings bok* ved 25 ars jubileet, 1935.

met and conversed while engaged in obtaining cargoes for their respective vessels.

The *Tacoma Sunday Ledger* commented on the movement of wheat by rail from U.S. Pacific Northwest wheat growing areas to Tacoma in a feature article for its readers:

> Grain is bought in the fields of Central and Eastern Washington, Idaho, Montana, and Oregon by local Tacoma grain buyers or by agents of the exporters. It comes to Tacoma in the early season at the rate of 80 to 100 rail car loads per day. It is generally sacked in the fields, although some of it comes in bulk, but exporters find it more satisfactory to ship in sacks. Doubtless the time will come when it can be just as safely handled in bulk.

Since no apparent effort was made in the field to place a specific weight of grain in each sack, bulk weighing of the sacked wheat became necessary either enroute or at dockside prior to loading. The British ship *Karoo*, which finished loading its wheat cargo just before the *Tamar E. Marshall* took its place at the dock, took aboard 51,245 bags totalling 6,860,480 pounds, an average of about 134 pounds per sack. No record of the number of bags put aboard the *Tamar E. Marshall* is available, so the average weight of bags in its cargo cannot be calculated.

Capacity of wheat sacks apparently varied from port to port. A resident of Clatsop County, Oregon, writing about his experiences as a shanghaied merchant seaman aboard the British bark *Musselcraig*, recalled watching native stevedores unloading sacked wheat (from a Columbia River port) into lighters alongside that vessel in Algoa Bay off Port Elizabeth, South Africa. He was particularly impressed that due to the weight of each sack—some 280 pounds—only one came out of the hold at a time, contributing to an unloading time of some three months.[8]

When a ship seeking a wheat cargo arrived in Tacoma its master arranged with one of the larger banks to honor his checks on account of his ship, paid off his men and paid off all other obligations that the ship owed. He then took on his cargo, hired a new crew or re-employed the old one, bought his stock of provisions and other supplies for his outward voyage, and settled with the bank by giving a draft on his owners payable at some European port. The wheat was paid for in the same way.

---

[8] Osborne, p. 7.

For more than 25 years prior to the recognition of Tacoma as the most available wheat shipping port on the Pacific coast, the rich grain fields of Eastern Washington and adjacent areas had sent their annual produce to Portland, Oregon for shipment abroad. But with the construction of ample wharf and warehouse facilities for grain handling in the young but ambitious city of Tacoma, a change occurred.

Prior to 1881, there had been no grain exported from Tacoma, Washington in commercial quantities. With the completion of the Northern Pacific Railway, many Tacoma businessmen were waiting for an opportunity to expand trade in any way possible.

Captain Isaac F. Gilkey of the American ship *Dakota* can probably be credited with Tacoma's being recognized as a major grain shipping port, even at Portland's expense, since that port had until then enjoyed a virtual monopoly on grain exports from the Pacific Northwest.

Having unloaded his cargo of railroad iron for the Northern Pacific Railway, Gilkey faced the practical necessity of obtaining a suitable cargo to carry away from Tacoma. To avoid the time-consuming trip around to the Columbia River and the high port charges being required at Portland, he explored the possibility of bringing a cargo of grain by rail either from Portland or from Portland's up-country sources to load into his empty vessel at Tacoma.

But after journeying to Portland and interviewing all the established grain dealers there, he could not find even one that would sell him a cargo for shipment to Tacoma for offshore loading at that port.

In spite of determined opposition from the Portland Chamber of Commerce and most of the merchants there, he finally found a single grain brokerage house, Balfour, Guthrie & Co., that would sell him a cargo on his terms. After protracted negotiations, he convinced officials of the Northern Pacific to haul his estimated 1,825 tons of wheat to Tacoma for $2 a ton, and to stagger the arrival of the 60 or more box cars on the dock tracks to be governed by his ability to complete loading of one day's supply before the next group of cars was placed.

Gilkey took his cargo of wheat (36,216 centals)[9], or 3,621,600 pounds, (valued at $51,000 by U.S. custom house documents) to Liverpool, where it was said that his vessel made the most profitable

[9] A name for one hundred pounds of a commodity, especially grain, suggested in 1859 by a lawyer named Danson, but not formally legalized for use in trade by the British government until about 1880.

trip of any in the world that year. Tacoma was on its way, and Balfour, Guthrie & Co. lost no time in opening an office there.

After the *Dakota*'s first wheat cargo from Tacoma, the traffic increased with each passing year. By the 1886-87 season, 11 vessels came for wheat, and by 1890 when the *Tamar E. Marshall* entered the port, she was one of 20 or more vessels in the "wheat fleet" that year. Warehouses, grain elevators and mechanical loading devices had long since taken the place of the laborious hand-loading of Gilkey's time, but even with those improvements, it was necessary for the *Tamar E. Marshall* to wait her turn to be loaded at the Warehouse Dock.

Towed down to Tacoma from Port Discovery partially ballasted, she "hauled in alongside" the *Karoo* on November 18 to take a "stiffening"[10] of wheat over that vessel's deck to enable her to discharge safely her remaining ballast out in the stream while waiting her turn at the dock. For about a week during that period, Captain James H. Parker, the long-time commander of the *Tamar E. Marshall* (other crew members who had not deserted on arrival at Port Townsend had been discharged) was driven to distraction by the intermittent but incessant sound of blasting aboard the nearby *Laura Madsen* and the barge *Ludlow*. These two vessels were attempting to unload cargoes of bituminous rock which had fused into an almost solid mass from heat and pressure in transit.

Finally, on November 24, 1890, the very day that the *Struan* left for Australia, the *Tamar E. Marshall* was towed to the dock and chutes were rigged for the gravity transfer of wheat sacks to the vessel. The speed of that operation depended almost entirely on how rapidly the sacks could be whisked away from the hatch area to the wings of the hold as they came down the chutes. Too few men would slow down the process while too many would just get in each other's way.

The basic minimum rate of loading was generally recognized at about 120 sacks per hour, per hatch. If three hatches were being loaded, then, some 360 sacks per hour could be put aboard. The bark took on her cargo in six days, indicating that there had been two shifts during each 24-hour period to keep the operation on schedule.

Sometime during early November, Parker had apparently requested that he be temporarily relieved of his duties. In response to that telegraphic message, Captain Andrew Freeman Kenney arrived by train from Saint John, N.B. in time to assume command of the

---

[10] A partial loading of cargo put aboard before all ballast has been unloaded, to make possible safe movement of the vessel from dock to dock or within the harbor for final loading after discharge of remaining ballast.

The derelict *Struan* stranded in front of the Chamberlain homestead on the Tillamook County, Oregon beach a mile south of Cape Lookout, just after dawn on Christmas day, 1890.
—*Original painting by Orvis Langland; owned by the author.*

Tamar E. Marshall on December 1st, after completion of loading, just prior to sailing, about a week after the *Struan* had departed for Australia.

At the published ocean freight rates—35s to 37s 6d per ton (U.S. $8.50 to $9 per gross ton of 2,240 lbs.) in effect when the *Tamar E. Marshall* reached Port Townsend for later Tacoma loading, the vessel would have earned gross freight charges of a minimum of almost $16,000 for transporting her 1,886 ton cargo of wheat to Sunderland—perhaps about one-third the value of the vessel itself—for its six-month involvement, including the time taken to load and discharge the cargo.

Within a relatively short time after permitting the *Tamar E. Marshall* to take a "stiffening" over her decks (to expedite loading where there was a shortage of dock space) the vessel *Karoo* was sued by members of its crew alleging failure to serve lime juice or other antiscorbutics for the prevention of scurvy on a protracted voyage. The court found that such omission was a breach of the ship's contract and justified the seamen leaving the vessel, though judicial notice was taken of the fact that the ship chandler who supplied the vessel at the beginning of the voyage was responsible, since the captain had expected to find the antiscorbutics among the supplies delivered for the trip.[11]

Deck boy Madsen's account of the voyage from Port Discovery summed up the beginning of the trip as a recurrence of the first part of the voyage from Liverpool—after just a few days, storms on all sides. A more detailed record of the *Struan*'s battle with the elements is found in the letter[12] Captain Skogland addressed to the owners in Haugesund on January 3, 1891 after his safe arrival in San Francisco the previous day, at which time he had dispatched a brief telegram advising them of the loss of his ship and the safety of all crew members. He wrote:

> We left Port Discovery on November 24, with a breeze from the South, more accurately from SE to SW, in a moderate to strong gale which lasted until the afternoon of December 2, at which point the wind increased to a storm gale from SE, with variations toward W. From 8 p.m. to 12 a.m. on December 2, the wind decreased somewhat and shifted toward NW. At 12 o'clock mid-

11 Johnson v. the *Karoo*, D.C. Wash. 1892, 49F 651.

12 An edited translation from the Norwegian newspaper *Haugesunderen*, January 31, 1891.

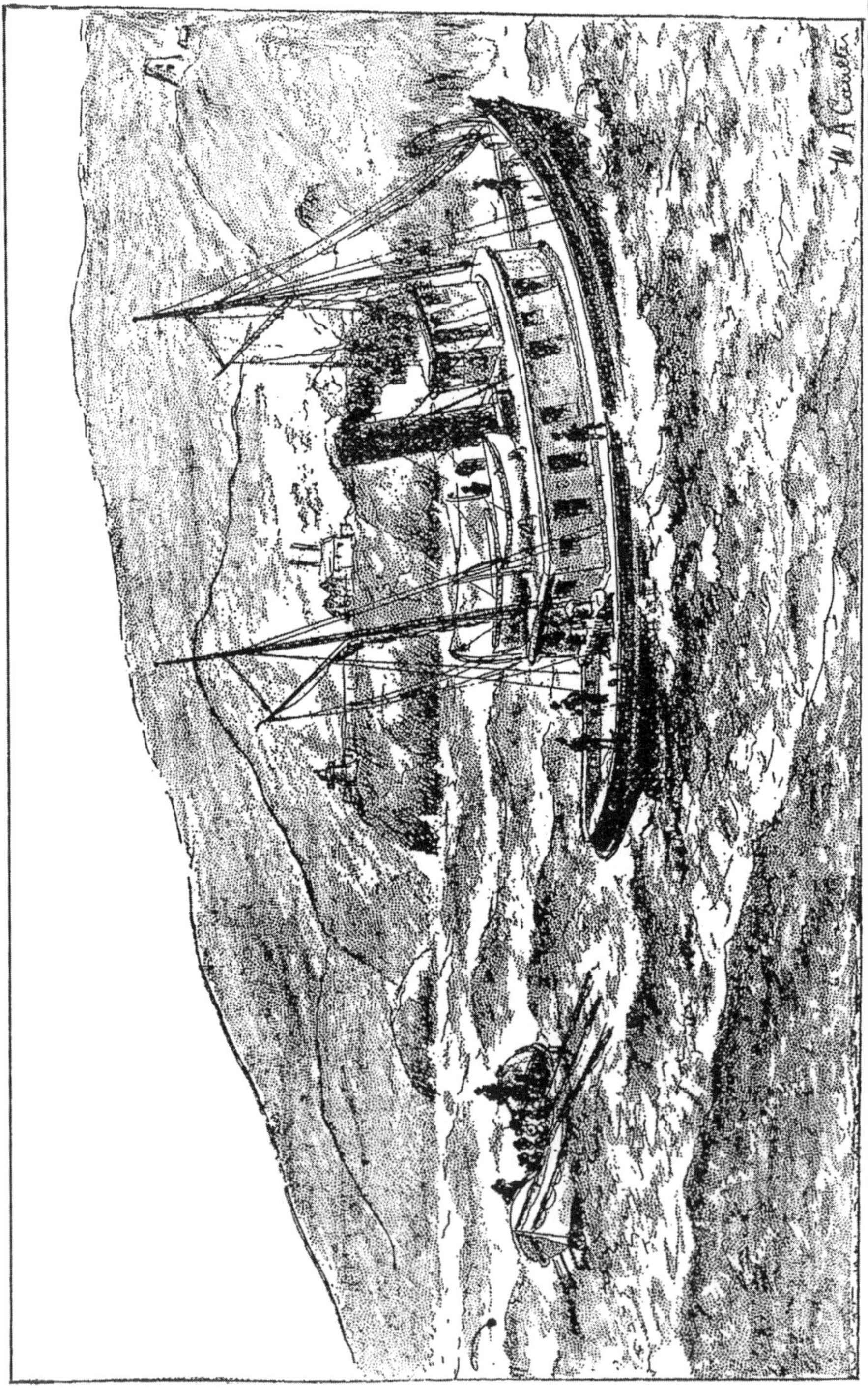

The Spreckels tug *Vigilant* and the life-saving crew from the Bakers Beach Station searching for the missing Whitehall boat and its owner off Point Bonita.

The lifeboat under way in the illustration opposite would have been very similar in size and design to the "great patched lifeboat" used to evacuate the *Struan* as described by deck boy Anton Th. Madsen in his later recollection of the incident.

Hundreds of almost identical all-purpose boats, designated as "standard 24-foot whaleboats," were in use by the U.S. navy until well into the 1920s.

Irish-American artist William A. Coulter, although employed by the San Francisco "Call" as a sketch artist, was also an accomplished and respected painter in oil of marine subjects, especially sailing vessels. Many of his larger and well-known works purchased by businesses and financial institutions for display in board rooms and public areas, were lost in the San Francisco fire of 1906. Some of his other paintings can be seen today in the Maine Maritime Museum, Bath, Maine; California Historical Society, San Francisco; and New York's India House.*

— Sketched by a "Call" artist. Courtesy National Maritime Museum Association, San Francisco.

* Dorothy E.R. Brewington, *Dictionary of Marine Artists* (Mystic, Connecticut: Mystic Seaport Museum, 1982) p. 95.

night I brought the ship sharply about, raised the lower top sails and steered SSE. I was personally on deck from 11 p.m. to 5 a.m. on the morning of December 3. The wind was fairly brisk, but the ship was coasting on small sails with the wind from aft, and she was sailing nicely. As previously indicated, I went below at 5 a.m. after having instructed the Second Mate to summon me in case of the slightest change. Between 7 and 8 a.m. I was awakened by the ship sharply listing and, arriving on deck, I discovered that the ship was lying leeward and all her sails had been blown away."

The Second Mate reported that hurricane gusts of wind had suddenly arisen which had tossed the ship in the direction the wind was blowing. The ship was now listing strongly to one side, and the sea was crashing over her, causing the cargo on deck to break loose. We attempted to get her under sail, but in vain. When we determined that it was impossible to right her, we heaved the deck cargo overboard (to leeward).

The windpump had been in steady operation, and I expected that the ship would be emptied of water at any moment. I sent the first mate to the hold to see how much water was in the ship, and he reported that it was above the lower deck. In the meantime, the hurricane intensified, and the water level increased in the hold, the windpump notwithstanding. At 3 p.m. the pipe broke. The ship was listing more and more to the side as she took on water, and we feared that she would capsize. We cut the topmast rigging at the crosstrees and the fore topgallant mast. When this proved to no avail, we cut the main standing rigging and, finally, the main mast itself.

The storm retained its severity during the night of December 3 to 4, and the sea was breaking over the ship as over a reef, partly damaging the deckhouse and cabin(s) and filling them with water. All the provisions which were not in tin cans or sealed containers were destroyed. The fresh water supply, which was in the lower part of the hold, was spoiled as well, leaving us quite helpless.

Under such conditions we remained on board for twelve days. On Sunday, December 14, we caught sight of a ship. The Third Mate, John M. Hauge, went to place a distress flag on the foretop but fell down and broke both thighbones. I applied splints to his legs in the most skillful manner possible, and it was with great difficulty that we managed to save him.

The ship which we saw had noticed our signals and approached us towards 6 p.m. It was not possible for us to get into our boats at that time, and the (other) ship kept sailing (next to us) the whole night. In the morning we climbed into one of our boats and, after much difficulty, came aboard the other ship. Its name was the *Tamar E. Marshall* from Digby, Nova Scotia.

While still on board the *Struan*, we had attempted to steer by means of rope while operating the emergency rudder, but all was in vain; she did not respond. As previously mentioned, we had no water, and the weather was not pleasant enough for us to get hold of any. We came closer to land every day; indeed, however, it was a coast where there are very few places to land in a boat even when the weather is good. As the crew wanted to leave the ship, I felt compelled to accompany them.

The ship which we boarded was destined for Sunderland, England, and it was with difficulty that we got the Captain to land us here (at San Francisco).

The *Struan* was seen on December 19, and on the 24th a tug captain tried to tow her but was then obliged to let go of her; he then stepped aboard and removed all the loose objects, such as instruments, maps, clothing, etc. Since then, nothing further is known about her.

His letter's poignant closing lines may have betrayed his doubts about the average Norwegian shipowner's belief in "going bare," or sailing without insurance.

I regret that we did not insure…
In haste, respectfully yours,
T. H. Skogland.

A few details from the deck boy's account serve to fill in facts the captain didn't consider important enough to mention in his initial report. Madsen recalled that prior to the rigging being cut away in an effort to right the ship, the yardarm[13] for the foresail was lying in the water, indicating that the vessel was very close to capsizing. He also reported that great damage was done by the heavy timbers carried on deck after the lashings began to break; one huge seventy- or eighty-foot long timber, he noted, was carried aft on the starboard side by the raging seas, smashed into the poop through the storeroom, broke

[13] The yardarm referred to was either end of the yard, the horizontal spar supporting the first sail above the deck on the foremast. That yard on the *Struan* would have been perhaps 25 or 30 feet longer than the width of the vessel, allowing it to project some 12 or 15 feet past either side of the ship. Its height above the deck was at least 30 feet, and above water level with the ship on an even keel about 35 feet or more.

down the captain's berth and came to rest with one end protruding more than ten feet past the stern frame.

The actual location at sea where the *Struan*'s sails were all blown to tatters and the mainmast was cut away in an effort to right the badly listing vessel cannot be established with any degree of certainty. At various times during December 1890, newspapers along the Northern California and Southern Oregon coasts had been reporting sightings by various mariners of a dismasted derelict in the area off Point St. George, about twenty miles south of the Oregon-California border. Although first observers thought the derelict to be some other vessel, a few reported it as the *Struan*. One shipmaster recalled that he had believed the derelict to be a new vessel because the metal covering on the hull had been shiny and bright.

The point at which the *Tamar E. Marshall* took aboard the *Struan* crew was established by Captain Kenney of the rescue vessel as latitude 42° 46'N and longitude 128° 50'W. Captain Skogland of the *Struan* confirmed in his conversation with newspaper reporters upon his arrival in San Francisco that they had abandoned his ship in about latitude 43°N. From approximately 41° off Point St. George to the location of the rescue at 42° 46' would have been about one hundred miles in a generally northward direction that the derelict *Struan* had drifted between December 4 and appearance of the *Tamar E. Marshall* on December 14.

Reading between the lines of Madsen's recollections, we can assume that much of the night of December 14 was spent in making a stretcher in which to transfer injured third mate Hauge. Madsen termed it a "box."

Although Skogland in his report to the owners stated that "it was not possible to get into our boats at that time," giving the impression that he felt it too dangerous to do so in the dark, with a rough sea running, it may very well have been that the *Struan* didn't have a seaworthy boat left aboard after its encounter with a series of storms which had damaged almost everything on deck.

Madsen's story states that "we had only a great patched lifeboat," leaving open to conjecture whether emergency repairs to it were frantically completed during the hours of darkness. From his description, we can easily conclude that the boat they used to evacuate the *Struan* was the largest boat carried, probably stored upside down in a rack above the deckhouse.

On finally boarding the *Tamar E. Marshall*, Captain Skogland insisted that the *Struan*'s lifeboat be towed behind the bark. Whether

Svend Knudsen Hoie, born December 8, 1850 at Avaldsnes near Haugesund, joined the navy at 17 years of age, but deserted from the king's ship *Baldur* at the port of Archangel in Russia on August 29, 1871. He did not return to Haugesund until five years later, at which time he signed as a seaman aboard the *Dronning Sophie* (Queen Sophie). On that vessel he made several voyages, interspersed with trips on other ships. He qualified as a mate on May 10, 1884 and sailed aboard the *Dronning Sophie* again to America and later to Scotland before joining the *Sigurd* as mate in June 1888 for service until August 1889. On November 8 of that year he signed on at Liverpool to be first mate on the *Struan*.

— *Photo reconstructed from an undated and unidentified newspaper clipping photo of the entire crew of the full-rigger Haugesund in 1900, at which time he commanded the ship.*

THE LOST STRUAN

The Crew of the Abandoned Ship Come Into This Port.

The Men Suffer Much Hardship--Captain Skogland Tells a Very Interesting Narrative.

The crew of the abandoned Norwegian ship Struan, for whose safety considerable anxiety has been often expressed since the news was received of her waterlogged condition, arrived here yesterday afternoon on the tug Vigilant. The first news of the rescue of the men was received here yesterday afternoon about 2 o'clock, when the Merchant's Exchange received information from their marine reporter at Meiggs' wharf that the Vigilant was passing down the bay having on board eighteen men, comprising officers and crew of the abandoned Norwegian ship Struan, which was bound from Port Discovery for Melbourne. The Vigilant took the men off the British bark Tamar E. Marshall, bound from Tacoma for Sunderland, inside the South Farallones, early yesterday morning. The Struan lost her rudder on the 3d of last month and the crew remained with the vessel until twelve days later, when, on account of her waterlogged condition, they left her and boarded the Marshall.

The news was posted at the Exchange and immediately quite a crowd of shipping men went down to the Oceanic dock to learn the particulars of the abandonment of the ship. The Vigilant arrived at the dock about an hour or so after the news was received at the Exchange, and from Captain McCoy, who is in command of the tug, an ALTA reporter learned that while she was outside the Heads early yesterday morning she was signaled by the British bark Tamar E. Marshall, which transferred to the tug the crew of the Struan. According to the statements which the crew of the Struan made to the reporter, it appears that they suffered much hardship. None were lost and only two were injured during their thrilling experiences. They had been on the Marshall since December 15th, when they were taken from the waterlogged craft.

Captain Skogland, who was in command of the Struan, tells a very interesting narrative of the terrible storm which the vessel experienced when only a few days out. "We left Port Discovery for Melbourne with 1,260,000 feet of lumber on November 24th," says the captain, "and three days later we encountered stormy weather. Soon the vessel sprung a leak and in spite of all that could be done the pumps became clogged. The vessel soon filled with water and on December 3d she became water-logged and unmanageable. The stores were kept out of water, but all the fresh water supply was lost, as well as clothing and medicines. From lack of water the men suffered intensely, but when it rained a little was caught and relieved our thirst somewhat. All the rigging possible and the mainmast were cut away to ease the vessel."

Captain Skogland then went on to say that during the gale Thomas Jorgensen, the second mate, was washed into the sea, but the captain succeeded in rescuing him. On Sunday, the 14th, a vessel was sighted, and John Hauge, the third mate, volunteered to take a signal flag to the foremast. At the crosstrees he lost his footing and fell to the deck, breaking both thighs. There was a heavy sea at the time. Captain Skogland set the fractures as best he could. The next day the vessel, which proved to be the Marshall, approached and took all hands off. They took with them one of the Struan's boats, and this was towed in to-day by the Vigilant. Aboard the Marshall they had every care, and all now appear little the worse for their experience. Hauge is doing very well. The Marshall was heading into this port, intending to put the men aboard a pilot boat, when the Vigilant met her and took the men off. The Marshall then proceeded on her voyage. Captain Skogland calculates the Struan was in about 43 N., 134° W., when abandoned.

The names of the entire crew of the Struan are as follows: Captain, T.H. Skogland; first mate, S. Hoie; second mate, Thomas Jorgensen; third mate, J. M. Hauge; carpenter, Johannes Olsen; cook, Thomas Thomassen; sailmaker, Ludvig Sernikov; sailors--Laurits Hansen, Ingvald Odland, Edvin Simonsen, Jacob Mortensen, Anton Th. Madsen, Ingvald Nilsen, John Andersson, Elias Trondsen, Gus Gjessen, T.A. Uhren, Petter Mattila.

What has become of the Struan is still a mystery, but the general opinion among shipping people seems to be that by this time she has gone to pieces. It will be remembered that the first news of the abandonment of the ship was received here from Captain Reed of the Ericsson. He did not know her name, and for a long time there was much doubt about her identity. Authentic information as to the name of the vessel was first known through Captain Johnson of the steamer Scotia, who ran across the Struan in her waterlogged condition and, with an eye to making salvage out of her, tried to tow her to port. While making fast to her the steamer Truckee came along, whose captain offered his assistance for half salvage. His offer was refused by Captain Johnson, who wanted all the prize or nothing. As a result, the strain proved too heavy a load for his craft, and the hawser parted. The Scotia then came on to this port and Captain Johnson made a report to the lumber company to which the Scotia is chartered.

The company considered the matter, and upon Captain Johnson expressing the belief that the waterlogged Struan was still afloat, or at any rate ashore, but not broken up, with her valuable cargo, it was decided to send the Scotia out again in search of the Struan. She left Thursday afternoon at a late hour with extra men and a full supply of apparatus to tow the Struan into port if she meets that craft. It is more than probable that success will attend the venture. Even if the Struan is ashore and breaking up, a good part of the cargo of white pine lumber can be picked up.

Reconstruction of an 1891 news article with corrections made in the spelling of names. *Daily Alta California*, January 3, 1891.

he had looked over that vessel's boats and calculated that there was not enough lifeboat capacity for both crews should some disaster befall the *Tamar E. Marshall*, or whether he contemplated using the boat to carry his own crew ashore if a suitable landing site should be discovered in good weather, is left to conjecture.

There is no doubt that if members of the *Struan* crew made repairs to their lifeboat after arrival of the *Tamar E. Marshall* on the scene, the strength and workmanship of that effort was of the highest quality. The boat survived being towed for 17 days at speeds sometimes far in excess of its design criteria. Its towline was transferred to the tug *Vigilant* which brought the boat into San Francisco harbor, along with the 18 shipwrecked crewmen.

One question has puzzled me since I first discovered details of the rescue at sea of members of the *Struan* crew. Why did the captain of the *Tamar E. Marshall* require the *Struan* sailors to use their own boat? Seventeen half-starved men (remember that third mate Hauge had two broken legs and according to Madsen's account, each seaman had been subsisting on a daily ration of a teaspoonful of canned milk and a few sea-water-soaked biscuits for ten days or more) were hardly equal to the task of manhandling the heavy boat (probably weighing about one thousand pounds) repairing, righting and launching it, climbing down an accommodation ladder to the lifeboat, lowering Hauge in his box, rowing across to the other ship and climbing up another rope ladder to the bark's deck. But they did it. It is no wonder that the *Struan* skipper related in his report to the owners that they had "climbed into one of their boats and, *after much difficulty*, came aboard the other vessel."

In fairness to Captain Kenney, some possible justification for his decision to require the *Struan* crewmen to use their own boat to board his vessel should be considered. He had been in command of the *Tamar E. Marshall* for only two weeks, having taken over from Captain James H. Parker, the regular long-time commander of the bark, the day he sailed from Tacoma. It is known from newspaper accounts at the time that he had no say in hiring any of his sailors, and that a "crew was put aboard the *Tamar E. Marshall*" by a Tacoma waterfront character who may or may not have always furnished willing, capable, experienced and sober seamen. Kenney had not had the time or opportunity to evaluate the individual capabili-

---

14 That description of sea conditions at the time of meeting the *Struan* was included in Captain Kenney's communication to Norwegian authorities after arrival at his destination, where he made clear that he expected compensation for the "deviation from course" to land the *Struan* seamen, and for the value of meals served them while aboard his vessel. See the Epilogue for results of Kenney's request.

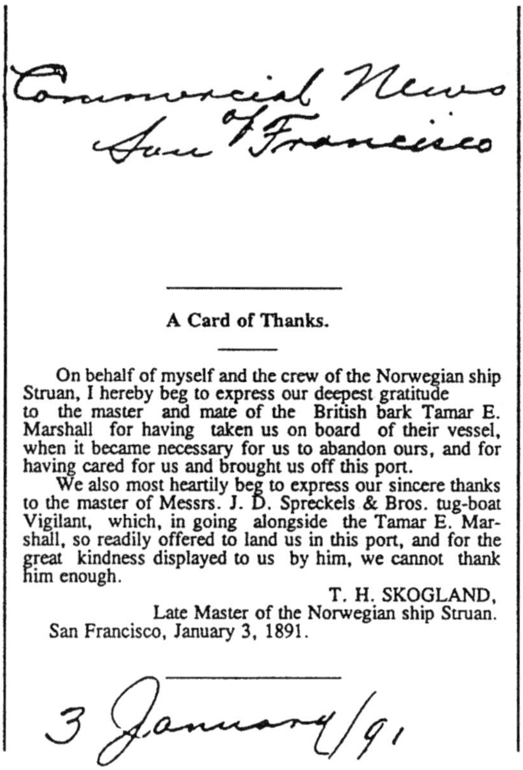

Commercial News of San Francisco

A Card of Thanks.

On behalf of myself and the crew of the Norwegian ship Struan, I hereby beg to express our deepest gratitude to the master and mate of the British bark Tamar E. Marshall for having taken us on board of their vessel, when it became necessary for us to abandon ours, and for having cared for us and brought us off this port.

We also most heartily beg to express our sincere thanks to the master of Messrs. J. D. Spreckels & Bros. tug-boat Vigilant, which, in going alongside the Tamar E. Marshall, so readily offered to land us in this port, and for the great kindness displayed to us by him, we cannot thank him enough.

T. H. SKOGLAND,
Late Master of the Norwegian ship Struan.
San Francisco, January 3, 1891.

3 January/91

ties of his men, and had no idea whether any of them were experienced boatmen. A very heavy sea was running at the time.[14] To have called out a boat crew under such circumstances might have resulted in loss of the boat and some or all of its crew, and necessitated putting in at some west coast port for replacement.

Captain Skogland, recalling that he had sent an open boat from his own storm-damaged ship while off the coast of France in a gale to take survivors off a sinking Dutch vessel a year earlier, may be excused if he found Captain Kenney's action less than friendly. In his report to his owners, Skogland's statement that "it was with difficulty that we got the Captain to land us here, in San Francisco,"suggests that a strained relationship existed between Skogland and Kenney while the *Struan* crew was aboard the *Tamar E. Marshall.*

For 17 days after taking aboard the *Struan* mariners, the *Tamar E. Marshall* sailed back and forth off the northern California coast seeking an inbound vessel to which the rescued seamen could be transferred. Captain Kenney had indicated that he wanted to sail directly for England with the *Struan* seamen, but realized that he did not have enough provisions aboard to feed the extra men on such a long voyage.

On December 29, the *Tamar E. Marshall* sighted the bark *George S. Homer* in prox latitude 37° 39'N. Captain Kenney stated in a later report that "we boarded that vessel and asked the master to take the shipwrecked crew (from us) as we had not provisions sufficient for the extra number of men for the whole voyage to England."

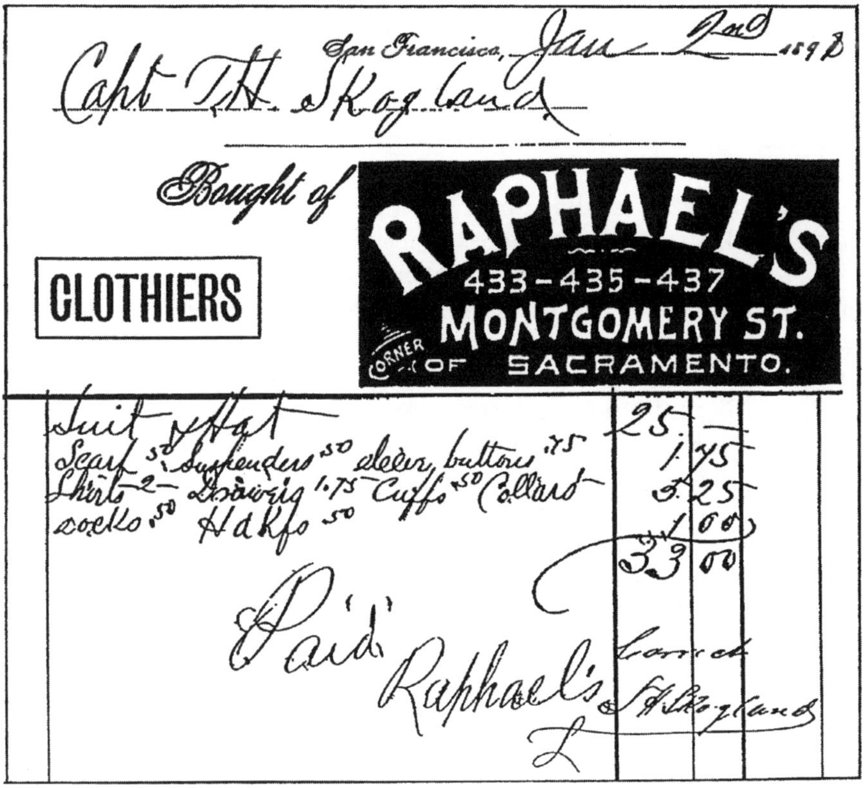

San Francisco, Jan 2nd 1898

Capt T.H. Skogland

Bought of RAPHAEL'S

CLOTHIERS

433-435-437 MONTGOMERY ST. CORNER OF SACRAMENTO.

Suit & Hat	25	
Scarf .50 Suspenders .50 sleeve buttons .75	1	75
Shirts 2— Drawers 1.75 Cuffs .50 Collars —	3	25
socks .50 Hdkfs .50	1	00
	33	00

Paid Raphael's L

Correct T.H. Skogland

Records of the Royal Norwegian Archives (Riksarkivet) in Oslo are so complete with regard to the loss of the *Struan* that it was able to produce even a copy of Captain Skogland's receipt for emergency purchase of replacement clothing the day he was landed in San Francisco.

Although the *Homer's* skipper refused to accept the *Struan* seamen, he did agree to sell some provisions to alleviate the shortage of food aboard the *Tamar E. Marshall.* Those extra foodstuffs, however, served also to strengthen Captain Kenney's determination to take the *Marshall* directly to England without stopping at any port to debark the *Struan* crew. It was indeed, as Captain Skogland advised in his letter to the owners in Haugesund, "with difficulty" that he persuaded Captain Kenney to "bear up" for San Francisco, then only about three or four days sail from the point of contact with the *Homer.*

Without the *Tamar E. Marshall's* logbook it is impossible to determine which boat was used to board the *George S. Homer.* It would be interesting to learn whether Captain Skogland volunteered the use of his boat (being towed behind the *Tamar E. Marshall*) and his boatmen (then rested and adequately fed after two weeks aboard the

Marshall) for that duty or whether Captain Kenney felt obligated to launch one of his own boats, now that the seas were calmer than those at the time he encountered the *Struan*.

Various newspapers in late November 1890 reported that there were more than fifty vessels loading or awaiting loading at Port Townsend, Discovery Bay and nearby Puget Sound ports while the *Tamar E. Marshall* was taking on wheat at Tacoma. A dozen of these craft were taking aboard either lumber or coal for the California ports of San Francisco, San Diego or San Pedro. Although the *Struan*, bound for Australia, and the *Tamar E. Marshall*, destined for England, would normally have sailed farther offshore than the California-bound vessels, it seems strange, considering the traffic along the mid- and lower-California coast that the *Tamar E. Marshall* did not sight any other vessels on its 17-day zigzag cruise between taking aboard the *Struan* crewmen and offloading them to the tug *Vigilant* near San Francisco.

The considerable space (almost an entire column) in one of the San Francisco papers devoted to the arrival there of the *Struan* crew on January 2, 1891 did not receive the wide attention accorded earlier reports of the shipwreck itself. Tillamook county residents near the actual wreck site who depended on mailed copies of the Portland *Oregonian*, did not know that all members of the crew had safely survived the *Struan*'s loss.

Captain T.H. Skogland of the *Struan*, on reaching San Francisco on January 2, 1891 with only the clothes on his back, sent the obligatory "thank you" notes to his rescuers and drew money from the Norwegian consulate to buy a new outfit while he awaited transfer of funds from Norway to pay off his crew. He dispatched a letter to the owners with details of the wreck, and returned to Norway to resume his career.

Chapter XVIII

John Fraser—The Shipbuilder

Shipbuilder John Fraser was the eldest of ten children. Born August 17, 1819, he had five sisters and four brothers. His parents were Alexander and Marion Maclean Fraser, who themselves were directly descended from John and Janet Matheson Fraser, the Pictou, N.S. Frasers who, after their arrival from Scotland, are believed to have been the first branch of the family established in the Canadian Maritimes.

Wright says John Fraser moved to the Saint John area probably in the early 1850s,[1] and that George Ambrose Fraser, his eldest son by his first wife, Pamelia McKenzie, was born there in August 1855.

In the 1860s, Fraser acquired what was then a parcel of rural land in Rothesay, New Brunswick. In 1870 he built a summer home there, and after the Great Fire of 1877 in Saint John, converted the property to a permanent residence.

The rigid apprenticeship system then prevalent in the British Isles for training craftsmen in various trades apparently did not survive the transfer to Canada of the hundreds of hardworking and frugal Scots who filled the immigrant vessels of the late 1700s. There are records of capable teenagers who hired out as tool carriers in the then-existing small shipyards to learn the trade; the only limit on their progression to craftsman status was the intelligence, initiative and determination they brought to work each day.

How or where John Fraser mastered the wooden shipbuilding business cannot be factually determined. It is believed that he had some experience in the Gabriel Merritt yard at Moss Glen (overgrown

[1] Esther Clark Wright, *Saint John Ships and Their Builders* (Wolfville, Nova Scotia, Canada: published by the author, 1976) p. 135.

with a one hundred-year old forest when I visited the site in 1991) and that he later joined the Nevins firm at Saint John, becoming master builder and a partner about 1864.

Almost from the time of his arrival in Saint John, Fraser had established his own and his family's connection with his church, St. David's Presbyterian. His generosity to that organization as his business prospered can be measured in his contribution of an elaborately carved lectern turned out by his cabinet makers and wood carver John Rogerson, and his support in the purchase of a bell for the church building when it was rebuilt after the great fire of 1877.

It is reported that after the original church bell had been used to help sound the alarm in the early hours of the Great Fire of June 20, 1877, it was cracked in its fall through the steeple as the building burned. A trio of active church members—John Fraser, James Shaw, and John Boyd—volunteered to donate a new bell. Fraser went to the Boston foundry of William Blake and Co., formerly Hooper and Co., to supervise preparation of the melt for the pour and to insure that proportions of component metals were maintained in accordance with specifications. After approving the mix he actually witnessed the pour itself. Later he arranged to have the new bell transported to Saint John on one of his vessels. The same church burned again in 1917; the bell was not re-installed after that fire but was stored in the boiler room until the 1980s when it was placed on the Sydney Street side of the lawn as a monument.

Fraser moved his residentce from Saint John to Rothesay (about eight miles apart) a year or so after the Great Fire of 1877, and soon became one of the leading figures in establishment of a St. David's Church in his new neighborhood. Family sources recall that he used the prestige of his shipbuilding business to encourage donations of lumber and materials by merchants in the area. He volunteered the services of shipyard employees to fabricate and erect the frames of the building, which turned out to be almost an exact architectural duplication of the St. Matthew's Church on Douglas Avenue, which in its turn had been built as a mission church by the original St. David's congregation in Saint John.

As many other successful shipbuilders of the time had done, the elder James Nevins established himself in Liverpool to better control the shipping business he had been building along with his ships. James Nevins, Jr. passed away in 1874, effectively terminating the Nevins-Fraser partnership. Upon settlement of the estate, John Fraser established his own business to continue the construction of vessels.

The greatest danger to a shipyard in the days of wooden hulls was from fire. The presence of boiling pitch-pots for caulking the seams between the planking, and the ongoing accumulation of wood chips and borings from the work of the adzemen and drillers created an ever-present hazard that could be alleviated only through extreme care, and constant removal of combustible scraps as they were produced. A bad fire in 1841 had destroyed numerous homes and business firms when a spilled pitch-pot ignited a shipbuilder's workshop. Fraser required that his laborers and apprentices clean up the yard and buildings almost as fast as scrap wood was generated. According to family recollections, and the comments of an old gentleman named Carpenter, with whom great-grandson Daryl Fraser often spoke as a young boy, the material was separated to set aside the larger stove-sized pieces to be distributed as fuel by his own employees to less fortunate families in the community.

In the ten years between 1876 and 1886, John Fraser launched 12 major sailing vessels—nine ships and three barks (spelled "barques" for Canadian and British readers), all built in his shipyard at the foot of Erin Street near the Marsh Creek bridge in Saint John, New Brunswick. Most were commissioned by various established shipowners, though he did construct two or three either exclusively or substantially for his own account. When building for others, he usually exercised the "builder's courtesy" of retaining a small fractional share in each of his creations, intending to insure for himself a continuing stream of income from the freights earned throughout the life of the vessel, at least during its early years or until the other owners had sufficient profits from their investment to buy out his shares. In his second vessel, for instance, the ship *Struan*, launched April 28, 1877, John Fraser's name appeared on the original registry certificate as the owner of eight shares—an 8/64th or 1/8th ownership.

A study of the Registry records pertaining to these vessels reveals that in spite of the fact several had masculine sounding names, 11 had female figures installed on their bows as figureheads. Only one, the *Struan*, launched in April 1877, carried a "three-quarter scale male figure." The small rural settlement of Struan in Perthshire, Scotland, was a stronghold of the Robertsons in the days of the clans, and it seems entirely possible that George Stymest Robertson, when he ordered the *Struan* built, may have specifically dictated the choice of some legendary clan leader as the subject for the ship's figurehead.

Some of the Fraser-built vessels were modified or remodeled a few years after their launching—in most cases to provide additional space for sail storage and larger captain's living quarters, etc.

Resulting decreases in registered tonnage ranged from 32 to 50 tons, as reported by surveyors' certificates originating in the ports of Liverpool, Cardiff, and Sydney, N.S.W., wherever the work was accomplished. No record of such renovations to the *Struan* has been discovered, though it is known that Captain James Cooper's family accompanied him on at least one extended voyage.

No one ever questioned the quality of John Fraser's ships. He always selected the very best materials,[2] hired capable and experienced workmen, installed the most up-to-date equipment available, and utilized the services of the finest specialists to carry on the finishing work his yard was not equipped to complete. His repeated sales of vessels to the Nevins fleet interests in Liverpool, the Troop fleet in Saint John and the Bennett Smith fleet operators and shipowners in Windsor all attested to the superiority of his ships, which were widely complimented in print, especially in the Saint John *Globe* whenever one was launched.

He built a comfortable home for his family and sent a daughter to school in New York, but managed to maintain a relatively frugal lifestyle. The very modest nameplate from the front door of his residence, engraved "Fraser" in plain script on a tiny 3 x 4 inch brass rectangle, passed to his great grandson Daryl Fraser, who displays it proudly on his own home in Saint John.

Some of Fraser's contemporaries might have considered him unlucky in his business dealings. A few would have attributed his bad luck in losing a number of his vessels to his disregard of an old Scottish sailor's belief that anything connected with death should not be taken aboard a ship. John Fraser ordered his gravesite marker from a firm in Scotland and is reported to have had it brought to Saint John on one of his own vessels, thereby attracting, according to the superstition, a string of bad luck.[3] That run of bad luck may have extended for fifty years after his death. Because the earth in the particular section of Fernhill Cemetery where he was laid to rest was of irregular contour, the foundation of his memorial column settled unevenly, and his descendants, when alerted by cemetery officials in the mid-20th century, had to remove the eight-foot monument until a new and heavier footing could be installed and the marker reset.

---

[2] Fraser's relationship with his suppliers is illustrated by the fact that a lumber surveyor was invited to take a minor ownership interest in two of his vessels and a timber merchant held a small share in the ship *Macedon*.

[3] James Clary, *Superstitions of the Sea* (Maritime History in Art, St. Clair, Michigan) p. 47.

Shipbuilder John Fraser's home in Rothesay, near Saint John, New Brunswick, as it appeared in 1990, no longer in Fraser ownership.

— Photo by Daryl and Lillian Fraser

Whatever the cause, his personal fortunes seemed to have plunged within just a few years after he launched his last ship in 1886. Having built 12 major vessels in a 10-year period, one of which he owned outright together with either substantial or minority interests in eight of the others, he found himself the victim of a series of marine disasters. Fires at sea or in port, collisions, stranding, storms, and disappearances at sea (failures to arrive) took such a toll that by the time of his death in 1901 his holdings had been reduced to include only very minor interests in but three vessels, and even those were counterclaimed by others.

The question arises as to whether the lost vessels had been adequately insured—or even if any insurance at all had been in effect in particular instances. It should be understood in this connection that Fraser himself could not be blamed in the event of no insurance coverage in some cases, since he held only a minority interest in a number of his ships, and the major or managing owner would have had the authority to decide whether or not to insure. Some maritime province shipowners were inclined to follow the Norwegian example. At least one such New Brunswick owner made no secret of the fact that he never paid a dime to an insurance company, and had never lost a vessel.[4]

Fraser's apparently strained relationship with his oldest son, George A. Fraser, born in 1855, is difficult to understand. Although George must have considered himself a partner with his father in construction of some of the vessels, the elder Fraser continued to refer to him as a clerk. Saint John city directories for the period involved list George A. Fraser as an occupant of John Fraser's residence and give his occupation as a clerk. On some of Fraser's ships, George appears in the registry records as an owner of a small fractional share, and in each such instance his occupation is again described as a clerk. All the original vessel registry documents list John Fraser's name alone as builder, and none recognize George as a joint builder, though the 1889-90 volume of Lloyd's Universal Register of Shipping shows *J. Fraser and Son* as builders of the *Struan*. There appears to be a possibility that George may have filed some legal claim to have himself recognized as a joint builder, since Lloyd's records normally reflected meticulously only information contained in official documents. What basis son George had for claiming shares of vessels remaining in John Fraser's possession at the time of his father's death can only be a matter for conjecture.

Fraser's first vessel built in his own name after severing on friendly terms a previous partnership with members of the shipowning Nevins family was the ship *Maritime Union*, launched June 24th, 1876. The 211 foot, 1,500 ton vessel was assigned official No. 72262 and signal letters SVKB. Charles Lewis Sanford was the first captain.

Original owners were:

Name	Residence	Occupation	Shares
James Nevins	Liverpool, England	Shipowner	20
John Fraser	Saint John, N.B.	Shipbuilder	32
James McGilligan	Saint John, N.B.	Stonecutter	2
Robert McGilligan	Saint John, N.B.	Stonecutter	2
William Lawton Prince	Saint John, N.B.	Builder	4
George Starboard Fisher	Saint John, N.B.	Patent roofer	4

[4] See pages 144 and 145, Chapter XV.

Finding stonecutters, a builder and a patent roofer in the owners' list leads us to assume that Fraser was trading interests in this first vessel for improvements to and additions at his Rothesay summer house as he gradually converted it to a permanent residence.

Disposition: Ship caught fire and burned at sea November 21, 1881 in 26° south latitude, west of South Africa.

The second ship launched from Fraser's yard was the 1,474 ton, 211 foot *Struan*, which went down the ways on April 28, 1877, to be assigned official No. 72293 and signal letters SVKQ. Captain James Cooper took command at launching and remained with the ship for more than eight years.

Original owners were:

Name	Residence	Occupation	Shares
George Stymest Robertson	Saint John, N.B.	Merchant	40
John Fraser	Saint John, N.B.	Shipbuilder	8
James Cooper	Saint John, N.B.	Master mariner	8
John Lane	Saint John, N.B.	Master mariner	8

Disposition: Ship seized and sold by the Lord High Marshal of the Admiralty Division of the High Court of Justice (United Kingdom) by order under seal dated 27 October, 1885, after the owners earlier that year did not pay when due an obligation resulting from a bottomry bond signed previously by Captain James Cooper at San Francisco to obtain funds to make the ship safe for the voyage back to England. Purchased supposedly at auction, by shipowner Andrew Gibson for a price criticized by some as far below the fair value of the then recently overhauled vessel. Involved in four incidents of grounding off the west coast of Florida in early 1887 while commanded by a captain who went insane. Transferred to Norwegian registry in the Fall of 1889. Dismasted and waterlogged in a series of storms off the west coast of the United States in December 1890. Crew members abandoned the ship for the safety of a passing vessel, and were landed at San Francisco on January 2, 1891, a week after the derelict timber-laden ship had drifted ashore and broken up on Christmas morning a mile south of Cape Lookout in Tillamook County, Oregon.

The 1,499 ton, 211 foot ship *Munster*, No. 72308, was Fraser's third vessel, sliding into Courtenay Bay August 9, 1877. John Lane was the first captain.

Original owners were:

Name	Residence	Occupation	Shares
William A. Robertson	Saint John, N.B.	Merchant	32
John Fraser	Saint John, N.B.	Shipbuilder	20
George A. Fraser	Saint John, N.B.	Clerk	4
John Lane	Saint John, N.B.	Master mariner	8

Disposition: Vessel sailed from Rio de Janeiro on May 10, 1895 for Newcastle, New South Wales, Australia in ballast, and was afterwards never heard of.

Fraser's fourth vessel, the 1,492 ton, 210 1/2 foot ship *Monrovia*, was launched May 18, 1878, and took official No. 72326 and signal letters SWFT. Captain Charles Lewis Sanford, who had taken the *Maritime Union* to England on its first trip, assumed command at launching, but was replaced by Captain William Johnson on June 15 before the *Monrovia* left Saint John on its initial voyage.

Original owners were:

Name	Residence	Occupation	Shares
James Nevins	Liverpool, England	Shipowners	32
John Fraser	Saint John, N.B.	Shipbuilder	16
Edwin Irvine Brass	Saint John, N.B.	Builder	8
William Lawton Prince	Saint John, N.B.	Builder	4
George Starboard Fisher	Saint John, N.B.	Patent roofer	4

Disposition: Ship sold to Liverpool owners in 1902, and reported transferred to Norwegian registry in 1905.

The *Macedon*, No. 72340, surveyed at 1,485 tons, was the fifth ship John Fraser built to his standard hull design. All five measured about 211 feet long and were about 39 feet in breadth (width). Only the vagaries resulting from differences in the characteristics of individual timbers used contributed to the very slight variation in length and width of the several vessels. Overall lengths ranged from a few inches over 211 feet down to the same number of inches under that measurement. Widths were even more closely controlled. Only an inch or two separated those measurements on all five ships. Again, Captain Charles L. Sanford commanded the craft from its launching on July 13, 1878 until he was relieved at Cardiff on February 13, 1880.

Original owners were:

Name	Residence	Occupation	Shares
James Nevins	Liverpool, England	Shipowner	44
John Fraser	Saint John, N.B.	Shipbuilder	16
Edwin Fisher	Portland, Saint John	Timber merchant[5]	4

Disposition: Vessel was converted into a hulk and moored in the Bay of Funchal (Madeira Islands). On December 9, 1905 she was run into by the steamship *Lugano* and sank at her moorings.

After his fifth ship, Fraser built two barks in a different design. The first of these, the *Norwood*, 1,197 tons, official No. 80007, (not to be confused with the 101-foot brigantine of the same name, built at Prince Edward Island, Canada, in 1873), launched August 30, 1879, was for the Troop fleet of Saint John, and measured just a few inches under 190 feet overall, and 38 feet in width. Captain William Odell Gregory made the first voyage in this vessel to Liverpool (probably with lumber and deals) where he was relieved by Captain Alexander Fraser,[6] born in 1844 at New Glasgow in Pictou County, Nova Scotia. He passed his examination and was issued Canadian Master's certificate No. 327 at Halifax in 1873.

He sailed as a mate on the *Komo*, the *Sacramento* and the *Wanderer*

[5] At the price of common lumber (prox $10 per thousand board feet) in effect at the time, a 4/64th interest in a vessel valued when launched at perhaps $64,000, would have involved value equivalent to some 400,000 fbm, several times the quantity required for construction of a single vessel, even if a few thousand feet had been diverted to improvement of the Rothesay residence. Fisher's investment may have related to lumber for a number of vessels.

[6] Believed to have been John Fraser's youngest brother, date of birth unknown.

at various times during 1874-75, and in 1876 made a three-month trip to North American Ports in command of the *Sacramento*. Later that year he was appointed master of the *Bonito*, (which broke adrift while in harbour in late April 1877).

In 1878-79 he served as short-term skipper of the *Nicosia*, then the *Cersia* and again the *Nicosia* (later wrecked on Sable Island in 1894) before being given command of the bark *Norwood* (John Fraser's sixth vessel). That assignment lasted until July 5, 1882, at which time the *Norwood* was wrecked on a reef on West Preparis Island in the Bay of Bengal. After returning to Saint John, he was made skipper of another of John Fraser's vessels (number three), the full-rigged ship *Munster*, where he served about a two-year stint.

After that service, there was a period of almost ten years in which he was not credited with sea duty as master of any vessel. In 1894 he commanded the *J.W. Holmes* on a short voyage to North American ports and the next year continued active duty as a shipmaster, taking over as skipper of the *Ulanda* for almost five years before being given his final command, the *Ocamo* in 1900. His date of death is not a matter of record in Lloyd's "Captains' Register."

Original owners were:

Name	Residence	Occupation	Shares
Howard Douglas Troop	Saint John, N.B.	Merchant	26
John Fraser	Saint John, N.B.	Shipbuilder	18
George Ambrose Fraser	Saint John, N.B.	Clerk	3
William Amos Honeywill	Saint John, N.B.	Sawfiler	2
Michael Oneil	Saint John, N.B.	Ship carpenter	2
Patrick Moran	Saint John, N.B.	Blacksmith	2
David Richardson Willet	Saint John, N.B.	Clerk	1
Olivia Jane McAfee	Saint John, N.B.	Wife of John McAfee	2
John Wesley Parker	New York, U.S.A.	Merchant	8

Disposition: Vessel was wrecked on a reef on West Preparis Island in the Bay of Bengal on July 5, 1882.

John Fraser built his second bark, the *Lynnwood*, vessel No. 80035, signal letters SWJD, from the same plans as the previous one, and entirely for his own account. Captain Michael Tucker took command at launching on July 24, 1880. He was relieved at New York on January 18, 1882 by Captain P. Sinclair. The port of registry was changed to Windsor, Nova Scotia on February 22, 1882 by the new owners, B. Smith and Company of that port.

Disposition: Vessel sold to Bennett Smith of Windsor, N.S. in 1882. Reported lost in 1896, no further information.

On May 14, 1881, Fraser launched another ship from the same general plan as his previous vessels of that class, the *Loanda*, vessel No. 80817. Although measuring out almost two feet shorter than the previous five ships, this vessel's tonnage was surveyed at 1,525 tons, some 25 tons greater than any of its sister ships, probably due to its slightly greater width. Alfred John Scott was the first commander.

In this ship, John Fraser, breaking his previous custom, took no ownership interest, the vessel having been built entirely for the Smiths.

Original owners were:

Name	Residence	Occupation	Shares
Bennett Smith	Windsor, Hants County, N.S.	Shipbuilder	38
Thomas Aylward	Windsor, Hants County, N.S.	Master mariner	8
Matthew Allison	Windsor, Hants County, N.S.	Farmer	4
Levi Smith	Windsor, Hants County, N.S.	Shipowner	4
Alfred John Scott	Windsor, Hants County, N.S.	Master mariner	2
I. Wesley Smith & Edmund G. Smith jointly	Halifax, N.S.	Merchants,	8

Disposition: Vessel wrecked on the island of St. Vincent on September 11, 1898.

Fraser didn't launch another vessel until September 1, 1882, when he sent the ship *Honolulu*, No. 85993, down the ways. That craft was ten feet longer and almost two feet wider than his previous ships, though that extra size translated into only one hundred extra tons on the registry, the vessel having been surveyed at just a hair under 1,600 net registered tons. John Fraser took the major ownership interest in this vessel for himself, reserving the minor holdings for members of his family, an employee, and a master mariner friend with whom he had enjoyed previous business dealings.

Original owners were:

Name	Residence	Occupation	Shares
John Fraser	Saint John, N.B.	Shipbuilder	40
Roderick Fraser	Saint John, N.B.	Ship carpenter[7]	2
George Ambrose Fraser	Saint John, N.B.	Clerk	8
John Fenwick Fraser	Saint John, N.B.	Clerk	4
Robert Wisely	Saint John, N.B.	Lumber surveyor	2
John Lane	Saint John, N.B.	Master mariner	8

Captain Hiram Edgett took this vessel to sea as the first of about a dozen officers who commanded her during the span of more than twenty years under the British flag.

Disposition: Ship sold to Italian owners July 26, 1902 (during the period in which John Fraser's estate was being settled in Saint John, N.B.) Reportedly renamed the *Adda*, broken up in 1913.

Fraser retained major ownership of his next ship, the *Honauwar*, 1,618 tons, vessel No. 88252, built from the same plans as the previous *Honolulu*, measuring some 221 1/2 feet in length. Launched September 3rd, 1883, she was at first skippered by Captain John Lane, who yielded command to Captain George A. Smith at New York at the end of January 1886, but took over again after only about three months.

Original owners were:

Name	Residence	Occupation	Shares
John Fraser	Saint John, N.B.	Shipbuilder	54
John Lane	Portland, Saint John, N.B.	Master mariner	8
Robert Wisely	Portland, Saint John, N.B.	Lumber surveyor	2

Disposition: Vessel caught fire and burned while in San Francisco harbor on August 20, 1889, and became a total loss.

[7] Believed to have been John Fraser's next-to-youngest brother, date of birth unknown.

Fraser's 1884 venture was an experiment in which he built for the Smiths of Windsor, N.S. a hull some 10 feet longer and 1 1/4 feet wider than his previous barks, giving the *Levuka* an overall length of just under 200 feet and a width of 39.7 feet. Registered with official No. 85450, the 1,425 net registered ton bark slid into the Courtenay Bay waters on June 10, 1884, to be initially commanded by Captain Theodore Harding Harris, who in effect played leap-frog with two other masters, and skippered the bark on three different occasions during the first 10 years of its existence.

Original owners were:

Name	Residence	Occupation	Shares
Bennett Smith	Windsor, Hants County, N.S.	Shipowner	26
Charles DeW. Smith	Windsor, Hants County, N.S.	Shipowner	8
John M. Smith	Windsor, Hants County, N.S.	Shipowner	8
Martin Gay Allison	Windsor, Hants County, N.S.	Merchant	4
Thomas Aylward	Windsor, Hants County, N.S.	Master mariner	4
Levi Smith	Windsor, Hants County, N.S.	Shipowner	4
John Mann	Burlington	Master mariner	4
Theodore H. Harris	Grand Pre, Kings County, N.S.	Master mariner	2
Edmund D. Smith	Halifax, N.S.	Merchant	4

Disposition: Vessel sold, while moored at Liverpool, England, to Italian subjects, on June 14, 1904, reportedly after having suffered an earlier shipwreck.

John Fraser's final contribution to the maritime world was his largest vessel—the 1,774 ton, 229 foot *Lucania*, a ship constructed for the Smith firm in Windsor, Nova Scotia. Launched on April 19, 1886, with No. 88394, signal letters KGLH, to be commanded by Robert Gibson, it was remarkable in that the registry certificate carried the name of the builder as John F. Fraser. Never before had a middle initial appeared with John Fraser's name on any official records. Even his will, written some 4 1/2 years after the launching of his last ship, was signed only "John Fraser." Similarly, the elaborate burial marker he pre-ordered omits any reference to a middle initial.

Original owners were:

Name	Residence	Occupation	Shares
Charles DeW. Smith and John M. Smith	Hants County, N.S.	Shipowners and joint owners	52
M. Gay Allison	Hants County, N.S.	Farmer	4
Robert Gibson	Hants County, N.S.	Master mariner	4
Edmund G. Smith	Halifax, N.S.	Merchant	4

Disposition: Vessel put into Simons Bay, Africa with cargo and ship on fire, where she was condemned. Wright dates that occurrence in *1886*, the same year the ship was launched.[8] Registry records, however, almost illegible from the newer information being written over the top of the already existing data, indicate that the ship's registry was not officially closed until August 22, 1893. Lloyd's Register of Shipping (1890-91) reports that this vessel was stripped of its old metal hull covering and recovered over new felt, with yellow metal (Muntz metal) in October 1889.

A slowdown in wooden vessel construction began in the early 1880s. Wallace[9] has estimated that there were still two thousand men employed in shipbuilding and the various supporting trades in the Saint John area as late as 1888, but that number was being rapidly reduced every year. As shippers and insurance carriers realized that iron and steel hulls built elsewhere were more watertight and therefore less subject to cargo damage, and as shipowners recognized the economies in maintenance and savings in operational costs that could be effected using metal hulls, orders for new wooden vessels fell precipitously. When shipowners discovered, after a vessel survey by any of the major classification agencies, that their older wooden vessels would be faced with major overhaul and repair costs to keep them insurable, many opted for a quick sale "as is" to foreign owners, especially the Norwegians. Progressive New Brunswick owners were buying iron or steel hulls as replacements for the timber vessels sold overseas.

[8] Wright, p. 136.

[9] Frederick William Wallace, *Wooden Ships and Iron Men* (Belleville, Ontario: Mika Publishing Co., 1976), p. 309.

John Fraser signed his four-page handwritten[10] will December 22, 1890, having present as witnesses his Barrister-at-Law, Ezekiel McLeod[11] and William A. Ewing, presumably an associate or employee in McLeod's office. In that document he provided that his estate should be administered by three trustees—his daughter Ella J. Fraser, his son John Fenwick Fraser and McLeod—for a period of four years after his death to provide for an orderly closing-out of his business interests and the sale of his "property both real and personal of every nature and kind wheresoever situate and howsoever described."

The trustees were directed to make, during that four-year period, from funds available in the estate, annual payments (to be disbursed in quarterly segments) to his second wife Louisa $750; to daughter Ella J. Fraser $200; to daughter Jessie A. Fraser $200; to niece Pamelia Fraser (daughter of his brother George William Fraser) $50; and to son John Fenwick Fraser, "upon whom the management of my estate will principally devolve" $400.

The will further provided for the sale from the estate at closing of such property as would be sufficient to provide for the purchase of securities or investments to guarantee that his second wife Louisa (Weldon Fraser) would receive a lifetime annuity of $750 per year, to be paid in quarterly installments, and that niece Pamelia Fraser should receive a one-time stipend of $500 before the rest and residue of the estate should be divided and distributed share and share alike to children George A. Fraser, John Fenwick Fraser, Ella J. Fraser, Jessie A. Fraser and William A. Fraser.

Trustees were authorized and empowered to pay son William A. Fraser during the said four years "so much money as they may think necessary or reasonable, but they shall not be compellable to pay him anything if they shall not think it necessary or reasonable."

Son William's apparent alienation from the John Fraser family may have resulted from his marriage to Teena Jemma Graham, an attractive Scottish lass sent to Canada under auspices of the Presbyterian church to work as a cook in a Rothesay home. Marriage to a "servant" or

---

10 John Fraser's will was not in his own handwriting, and was presumably prepared beforehand by a scrivener in McLeod's office.

On the date of signing, Fraser's second vessel, the full-rigger *Struan*, then operating under the Norwegian flag, was a dismasted derelict abandoned by its crew seven days earlier some 60 miles off the Pacific Coast of Oregon. There is no evidence, however, to indicate that news of the loss of another of the ships he had constructed reached Saint John until after the safe arrival of the *Struan* crew at San Francisco in early January 1891.

11 Ezekiel McLeod was the public trustee at the time, and later became a member of the Supreme Court of New Brunswick.

"hired hand" in those days was in some circles considered a "no-no." Living descendants of the shipbuilder recall that John's daughter Jessie treated "Jemma," as she was known, with a degree of aloofness amounting almost to disdain, though Jemma always treated Jessie with "kindness and dignity" in return.

Fraser's will "authorized and empowered"[12] his executor and executrix (collectively known as the "executors" under Canadian law)[13] to insure and keep insured "all or any of my vessels or other property in such amount or amounts as they may deem best, out of the funds of the said estate."

Upon his death some ten years later (on January 27, 1901) an inventory of John Fraser's property by the two remaining trustees (son John Fenwick Fraser had predeceased his father) listed *no real estate*. It must therefore be assumed that any such real property he had held was disposed of prior to his demise.

Other property included:

> Six shares of the ship *Macedon* (launched in 1878), three of which were in the name of John Fraser and three in the name of Robert Thomson. (Two of these shares were claimed by George A. Fraser as his property)
>
> Valued at: $900.00
>
> Six shares of the ship *Monrovia* (launched in 1878), three of which were in the name of John Fraser and three in the name of Robert Thomson.
>
> Valued at: $450.00
>
> Four shares in the ship *Honolulu* (launched in 1882), one of which was in the name of John Fraser and three in the name of John H. Thomson. (Two of these shares were claimed by George A. Fraser as his property)
>
> Valued at: $600.00

The estate inventory references to Robert Thomson and John H. Thomson in connection with liens or encumbrances against shares of vessels held by John Fraser at the time of his death are presumed to

---

12 "Authorized and empowered" leaves the decision whether to do it to the trustees themselves. Had he wanted to make certain that such insurance was kept in place, he might have substituted "directed."

13 Thomas Hutchinson Tristram, W.F.L. De Quetteville, and Bernard H.H. Thomson, *Coote's Common Form Practice and Tristram's Contentious Practice* (Canadian Fourteenth Edition by A.C. Forster Boulton, M.P., Toronto, Ontario: The Canada Law Book Co., 1970) p. 29.

be to the two sons of William Thomson, founder of the firm of William Thomson & Co. of Saint John. William, born in Dundee, Scotland and brought to Saint John as a child by his parents, began as a shipbroker in a small way about 1848, and by the early 1860s had become owner of a modest fleet of smaller, locally-built barks and other vessels. The firm subsequently acquired substantial holdings in major carriers, including 17 sailing vessels and nine iron steamers.

William Thomson was one of the first Saint John-based shipowners to recognize the advantages of the iron ships being constructed in Scottish shipyards on the River Clyde and elsewhere even before the early 1860s. He had sent his son Robert to Liverpool to be trained as a clerk, and upon his return Robert was able to persuade his father to invest in more iron vessels in spite of their greater initial cost. Pointing out that his experience in British shipping offices had made him acutely aware of the greatly reduced maintenance costs for iron vessels, Robert influenced his father's vessel purchases to the point where, by the time William retired and turned the business over to his sons in 1882, a significant portion of the fleet consisted of iron hulls.

It became evident soon after filing of the will in probate court that shipbuilder John Fraser's business and financial affairs at the time of his death were substantially reduced from conditions in effect when he had signed his will ten years earlier.

Five sworn claims representing judgments against John Fraser during his lifetime, and against his estate, totalled more than $3,000. Sworn claims of "contract creditors" as the court defined them, came in at over $1,400, and there were unsworn claims which, when combined, equalled less than $200.

Based on this preliminary information, executor Ezekiel McLeod, on February 20, 1901, in his petition for letters testamentary, advised the Kings County probate court that the estate of the deceased "does not exceed one thousand dollars." But William A. Ewing, the other witness to Fraser's will, acting as proctor[14] for the executor and executrix (the executors) began a thorough, imaginative, and aggressive effort to straighten out the affairs of the estate. Keeping in mind the precepts put forth in the law that "in any probate action the function of the court is not only to do justice between the [remaining] parties, but also to do justice to the deceased by ascertaining and ultimately by its decree giving effect to all duly executed testamentary instruments...available...,"[15] Ewing discovered that a single judgment

---

14 A person appointed or employed to act in another's behalf for a civil court action.
15 Tristram, *et al.*, p. 507.

had been mistakenly entered twice, against two of the vessels in which Fraser had held shares. He accepted $150 from William Thomson & Co., and yielded two shares in the ship *Honolulu* to effect the release of all claims against the shares in the other vessels held by the estate, and compromised a $782 insurance company judgment by paying $100 to the firm's liquidators.

In one instance, proctor Ewing found himself arrayed in court against five or six barristers[16] representing various claimants to the estate assets. By reporting on the paucity of funds in the estate to satisfy all claims, both judgment and contract, the proctor, perhaps unintentionally, planted the seed for disagreement among the various counsel present. One barrister, appearing for a judgment creditor, requested and was granted a five-week continuance in order that he might prepare to argue whether a judgment creditor should take precedence over a contract creditor.

When the hearing reconvened, however, none of the creditors' representatives was present. It might be safe to assume that all the claimants, having learned of the relatively insignificant size of the estate and realizing that there were not sufficient funds to cover their claims, decided to save the cost of additional legal representation.

Over a period of several months, Ewing negotiated with bankers, the Registrar of Shipping, claimants' representatives and others, to the extent that by the time he had cleaned up the titles to the ship shares John Fraser had held at death, executor Ezekiel McLeod was able to sell the remaining shares at auction for over $2,550. At the February 21, 1903 hearing when the "accounts were passed" by probate judge George G. Gilbert, the court found the executor and executrix "indebted to the estate in the amount of $2,260.63, to be by them expended in manner required by law." The records we have been able to locate, unfortunately, do not detail how those funds were actually distributed. Whether the claimants against the estate or John Fraser's devisees received the money, and in what proportion, we have not been able to learn.

Family members recall that John Fraser's daughters, Jessie and Ella, following his death in 1901, lived well as "ladies," travelled extensively, and "never worked a day in their lives." Although it seems understandable that their father had transferred ownership of the

[16] Under British law, a counselor or advocate admitted to plead at the bar of the higher courts of law in protection and defense of clients, as differentiated from solicitors, who are permitted to advise clients and plead cases in lower courts, and to prepare cases for barristers to plead in the higher courts.

Rothesay home to the girls well before his own death, there appears to be some question as to whether other funds may have been similarly made available. Daryl Fraser (John's great grandson) remembers as a boy being told that "great Aunt Jessie was living on the interest from her inheritance" (which must have been substantial to have kept her in comfortable circumstances for more than 50 years). But on Jessie's death in 1952, Daryl's father, while paying her last expenses, discovered information leading him to believe that her inheritance had been exhausted under a cloud of suspicion, and that her funds may have been mismangaged or carelessly dissipated by her financial or legal advisors over a long period of time.

Chapter XIX

100 Years Later—
The *Struan* Shipwreck Centennial

Early in the 1980s there had been some talk among old-time scouts and scouters at Camp Meriwether about doing something to commemorate the 1890 wreck of the full-rigger *Struan*, whose anchors and anchor chains were so familiar to the 60,000 or more campers who had spent time there.

I'm not exactly certain when the decision was actually made to go ahead with plans for a 100-year remembrance of the *Struan*'s loss. There were several conversations about the idea with Lloyd Lillie, who had been the field executive for sea scout activities in the Portland area and popular waterfront director at Camp Meriwether for most of the 1930s. One of his responsibilities in that position had been to supervise the surf dips in the early days at the camp, where scouts using the buddy system romped in the ocean at the very site where the *Struan* had come ashore on Christmas day in 1890. Some of the discoveries made during those early surf bathing sessions inevitably led to the conclusion that more parts of the vessel were in the sand at the wreck site awaiting recovery.

We broached the idea for a Centennial celebration to Kenneth A. Wells, who had been present as a young scout leader with "Chief" Oberteuffer and other officials of the Portland Area Council of the Boy Scouts of America when they first inspected the abandoned Tillamook County homestead (the future home of Camp Meriwether) with the Chamberlain heirs in 1926. Ken had also worked as a field executive in the Portland scout office and had been a leader and camp quartermaster at Meriwether for many years. He had later traveled in Europe and was acquainted with some of the executives of the World Scout organization. Through his efforts, we were able to identify and approach a scoutmaster in Haugesund,

Original Centennial Artwork by Marianne Hauge, Bergen

Harry Stromo, boy scout leader, accordionist, trombonist, band director and head custodian at a Haugesund, Norway school, holds aloft a link of the ship's anchor chain presented to him at the *Struan* Shipwreck Centennial Celebration at Camp Meriwether in 1990. Note the finger-thin diameter of the upper part of the link after 100 years of exposure to the elements. The original diameter of a *Struan* chain link iron (wire size) was a full 1 5/8 inches.
— *1990 photo by Joan Crichton*

Norway (where the *Struan* had been registered at the time of its loss).

Harry Stromo, long active in the Boy Scout troop sponsored by the Salvation Army in Haugesund and as a band instructor to school children there, was selected as the first contact. When Stromo realized that our request for information about the *Struan* would necessitate considerable correspondence in English (we had no local volunteer Norwegian translators at that time), he pleaded his inexperience in our native tongue and prevailed on another Haugesund scout leader, Johan Eide, to take over the "search and report" duties. Eide had also been scoutmaster of the Haugesund troop and was conversant in English to a degree that enabled him to write letters in that language without noticeable difficulty, and to understand requests for specific information posed by questions from Portland.

It was soon realized that Wells, living on the east coast, was a whole continent away from what was going on in Portland, Oregon in regard to the contemplated event, and Wells asked Eide to address future correspondence on the subject to this author.

As a sort of combination investigator, researcher, detec-

The Haugesund Scout Band

Band members were all scouts or leaders in the Haugesund scout troop sponsored by the Salvation Army.

Trygve Togersen, Bandmaster	cornet
Vegard Togersen	cornet
Reidun Fossland	cornet
Sten Arne Eide	horn
Dan Jorgen Eide	horn
Harry Stromo	trombone
Audun Hansen	bass
Johan Eide	baritone
Tormod Togersen	euphonium

---

tive and good will representative, Eide, for about two-and-a-half years in 1988-89 and early 1990 covered a large section of southwest Norway in his search for information about the *Struan*. He drove around the countryside interviewing possible descendants of *Struan* crew members. He interceded with the Royal Norwegian Archives (the Riksarkivet) in Oslo to expedite my earlier request for details from that source. His detective work developed a lead to the identity of the descendants of the *Struan*'s third mate, John Mandius Hauge, and Eide journeyed to Bergen to find them and acquaint them with our plans for a centennial recognition of the vessel's loss. Somehow he cajoled one hundred-year old family photographs from families once connected with the ship. He dug out from old newspaper files Captain Skogland's original report to the owners about the wreck. His easy rapport with newspaper reporters led to a full-page illustrated article in *Haugesunds Avis* about the *Struan* and our plans for a centennial celebration. Whatever information we thought might be helpful, he found and sent to us.

Professor Tom Birnie of Portland State University was recruited as a volunteer translator for the many pages of old records and family histories that Eide discovered in Norway and sent over as background information.

One of the photos included with a letter from Eide in early 1989 was a picture of a young boy scout—Eide himself at about 12 years of

ANNOUNCING THE

STRUAN

SHIPWRECK CENTENNIAL AT CAMP MERIWETHER
SATURDAY, AUGUST 11, 1990 — 10:00 AM TO 4:00 PM

Meriwether campers all remember the rusting anchors from the *Struan* wreck around the flagpole, the name board hanging in the camp messhall, and the stories told around so many campfires of the Norwegian full-rigger stranding on Christmas morning in 1890 right in front of the Chamberlain homestead on the meadow.

Now, almost 100 years later, after a two-year long search, direct descendants of an officer on the *Struan* at the time of the wreck have been located in Norway and have agreed to be present at our centennial.

Come to the celebration! Meet with and talk to the grandson and great-grandson of the 22-year old third mate on the *Struan.* Read the diary written by the deck-boy on the last voyage. See Captain Skogland's actual report to the owners about the wreck with the poignant last line — "I wish we had insured . . ."

Browse through an extensive display of pertinent maritime records including the original ship's registry papers, crew lists, consular endorsements, old newspaper clippings and photographs, etc., provided by museums, universities, government archives, historical societies and related organizations throughout North America and Europe.

Hear details of the rescue of the entire *Struan* crew by the great grandson of the builder and major owner of the rescue vessel *Tamar E. Marshall* of Digby, Nova Scotia, Canada.

Listen to the scouts in a brass band from the *Struan's* home port, Haugesund, as they entertain during the day and as they play the Norwegian national anthem while their flag is raised over the wreck site and the international code flags of the *Struan* and the rescue vessel snap in the breeze from the signal halyards.

Visit the special *Camp Meriwether Station* officially sanctioned by the US Postal Service for the day of the celebration, where a special pictorial cancellation commemorating the shipwreck will be applied to all properly stamped envelopes.

A photo session with the international visitors will be arranged for camera fans. Bring plenty of film.

Your noon meal, consisting of "skibbruddsrasjoner med picnic-lunsj" is included in the registration fee and will be served to all registrants during the celebration.

To help alleviate a tight parking situation, restroom-equipped luxury motor coaches will be operated for participants who do not wish to drive their own cars. Coaches will discharge passengers right at the mess hall, making any travel from parking areas unnecessary.

For those who drive their own vehicles, transportation by private van from outlying parking areas to the general camp area will be available.

Come and join us on August 11, 1990. Send in your registration form today! This won't happen again for a hundred years!

Duane Rhodes
Asst. Scout Executive

Whit Crichton
General Chairman

A *Struan* anchor and chain as seen during the Centennial celebration of the shipwreck in 1990. Flags on the signal halyards are the call letters of the *Struan* (SVKQ) on the right and the *Tamar E. Marshall* (JBHL) on the left.
— *Photo by Joan Crichton*

Struan deck boy Anton Th. Madsen with his wife and son, John Tilmar Madsen, about 1912. Seated in the wicker chair at the left is John Enerstvedt. On the right next to Mrs. Madsen is Teodora Enerstvedt. The Enerstvedt children and their mother, who was Mrs. Madsen's sister, lived with the Madsens after the father was lost in a shipwreck in the Mediterranean.

—*Photo courtesy of Johan Enerstvedt's daughter, Randi.*

age—resting his band instrument on his knees after a music session at his local scout camp.

"How about coming to our centennial with a group of your scout musicians to play the Norwegian national anthem while your flag is raised at our celebration?" I asked. "We don't have any funds for travel expenses," I hastened to explain, "but we'll put you up with host families and you'll never have to spend a dime while you're here."

Everything seemed to come together. An anonymous donor here sent 20,000 Norwegian kroner to help with air fares, the mayor of Haugesund found some funds to make available, and John Hauge III in Bergen persuaded one of the shipping companies there to make a contribution to the cause.

The *Struan* Shipwreck Centennial was scheduled for August 11, 1990. There had been some earlier talk of holding it on the anniversary of the actual stranding (December 25th) but that idea was abandoned almost immediately when it was realized that ice on the highways at that time of year might have prevented anyone getting to the site; that many who might otherwise have attended would not care to interrupt family activities during the Christmas holidays; and finally, that the Norwegian scouts, if they came, had to come during the summer vacation when school was not in session.

Once the Haugesund scout group had committed to attending the *Struan* Centennial Celebration at the wreck site, preparations for that event were shifted into high gear, or in the vernacular of the *Struan*'s time, "they sailed freely before a following wind."

Invitations to host the group began to pour in. At the request of the visitors-to-be themselves, arrangements were made for the Norwegian musicians to give band concerts at senior citizen centers in the Portland area. Invitations to picnics and barbecues materialized.

By the time the Haugesund scout group arrived at the Portland International Air Terminal (PDX), the calendar of events for their entire stay was practically filled. Johan Eide (and Harry Stromo, too) with seven other Norwegian scouts and scouters, arrived in Portland on the afternoon of July 26, 1990.

A few days later, John Hauge III, grandson of the *Struan*'s third mate at the time of its loss, arrived from Bergen with his wife, son, and daughter and the son's fiancee—all to be serenaded on arrival by the nine-piece band from Haugesund with lively welcoming tunes and the Norwegian National Anthem. From all the hoopla taking place, some of the other passengers in the arrival lounge, seeing the Norwegian

The present-day Sandlake Country Inn was built as a private residence during the two years immediately following the *Struan* shipwreck. In this picture, timbers from the ship's cargo, long covered with wallpaper, have been exposed to show how the 4" x 10" and 4" x 12" timbers were utilized in the construction.

— 1990 photo by Margo Underwood

flag being displayed, must have thought the Norwegian king himself was arriving.

Knowing of His Majesty's long-time close association with the Norwegian scouting organization, and fully aware of his previous visit to Oregon for a ski vacation while he was crown prince, Portland organizers of the Centennial had invited his personal cooperation in the celebration and were severely disappointed by the tone of a letter from the royal secretary informing them that the Norwegian monarch could take no part in the celebration, not even send a greeting. Reasons for, and justification of, this seemingly uncooperative message became clear when it was later revealed that His Majesty, even before receiving the invitation to be part of the celebration, was confined with what turned out to be a terminal illness.

After a few days in Portland for sightseeing and filling their concert commitments, the Haugesund scouts, including the lone female member of the contingent, and their leaders were welcomed at Camp Meriwether for a three- or four-day experience with their U.S. counterparts. The Norwegian guests at mealtime were seated in the dining hall, one at each table, to encourage conversation, and rotated table locations at each meal so that by the time of the actual Centennial itself every scout in the camp had enjoyed an opportunity to talk with each guest.

On the very morning that the visiting Norwegians were scheduled to arrive at Camp Meriwether, scout Nathan Shult, while walking in the wet sand at the edge of the ocean directly in front of camp headquarters, almost tripped over a previously undiscovered eight- or nine-link section of the *Struan* anchor chain. He dragged his find back to the lodge area and later very generously agreed to allow his discovery to be separated in order to provide a single link for presentation to each foreign guest as a souvenir of the coming celebration.

The Hauge party enjoyed the hospitality of Margo Underwood and her husband at the Sandlake Country Inn, built as a farm residence between 1892 and 1894 by homesteader W.C. King and his son, largely from lumber and timbers spilled onto the beach when the *Struan* broke up in the surf, and hauled three miles to the building site in horse-drawn wagons. The Hauges were also present at a reception for about 50 invited guests from the Tillamook County area at the Inn on the day before the Centennial celebration at Meriwether.

At low tide, the Norwegians were taken on a tour of the actual area where the *Struan* grounded on Christmas morning 1890, and from which the anchors, chain, windlass and capstan were subsequently recovered.

Dr. Clarke Fraser (left) and John Mandius Hauge III (right) meet in front of the flagpole at Camp Meriwether during the *Struan* Shipwreck Centennial Celebration. Fraser is the great grandson of the builder of the *Tamar E. Marshall* which rescued the *Struan* crew at sea, and Hauge the grandson of the third mate aboard the *Struan* at the time of that vessel's loss. At the extreme left is the 16-inch diameter iron ball used to keep the *Struan* anchor chain close to the bottom for better holding.

— *Photo by an unidentified passerby who took the picture with Dr. Fraser's camera.*

The *Struan* Shipwreck Centennial celebration opened at Camp Meriwether on the morning of August 11, 1990 with an impressive flag raising ceremony during which the national flags of the three countries involved were run to the top of identical poles while the Haugesund scout band played the appropriate national anthem as each flag was hoisted. After the "Star Spangled Banner" for the U.S. Stars and Stripes, the Norwegian colors went up to the strains of "Ja vi elsker dette landet" and the Canadian Maple Leaf ensign rose to "Oh, Canada."

Introduction of the various international guests was followed by a scheduled photo session and visitation period during which everyone present was encouraged to mingle and talk with the foreign visitors.

The Haugesund scout musicians, led by bandmaster Trygve Togersen, played a lunch-time concert while those persons gathered for the event ate box lunches, each containing, in addition to sandwiches and fruit, a ship's biscuit or pilot cracker similar to those which provided the only food available to the *Struan* crew members during their 10-day stay aboard their derelict vessel in December 1890. Someone suggested that those persons desiring to add a touch of realism could replicate conditions aboard the ship at that time by soaking their biscuits in sea water. It is not recorded that anyone actually followed that suggestion.

Daryl Fraser, great grandson of John Fraser, builder of the *Struan*, flew in for the Shipwreck Centennial from Saint John, New Brunswick, Canada with his wife Lillian and entertained all comers with a colorful description of his teetotalling great grandfather trying to keep his shipbuilders, on their way home from a vessel launching, from visiting and patronizing every tavern on the route—one on each corner at some intersections in old Saint John.

Dr. Clarke Fraser of Montreal, P.Q., Canada, great grandson of Alpheus Marshall, builder and a major owner of the *Tamar E. Marshall* (which vessel, named for his wife, rescued the *Struan* crew members from their derelict ship at sea) recalled as a child watching ox carts pass the Marshall home in Bear River, Nova Scotia, originally built by Captain James F. Parker, long-time commander of the *Tamar E. Marshall.*

John M. Hauge III, completely at ease with the English language, reviewed the shipping careers of the two John Hauges born since the *Struan*'s time, gave some insight into conditions aboard sailing vessels before the turn of the century, and compared the size of the *Struan* with currently operating cargo vessels.

Robert Chamberlain exhibited the rifle used by the original Chamberlain homesteaders, a pair of field glasses found on the beach after the wreck, and the original six-lever padlock from the Chamberlain mail box, which, more than one hundred years before, had been mounted on a tree three miles from the homestead beside a wagon road partially maintained by the Chamberlains and other settlers in the area working off their county tax obligations.

An extensive display of photographs and old documents pertaining to the *Struan* from the time of its launching in 1877 through the safe arrival of the rescued crew at San Francisco in early 1891 was available for inspection during the entire day.

The U.S. Postal Service operated a temporary office at Camp Meriwether for the day of the celebration only, where a special postmark commemorating the *Struan* Shipwreck Centennial was affixed to all properly stamped letters or cards, which could then either be retained by the mailing party or deposited in the U.S. mail at the mobile station for delivery without further cancellation.

During the closing ceremonies, mementos of the event were presented to the Norwegian and Canadian visitors. Each received an actual link of the *Struan*'s anchor chain, while the international code flags designating the call letters of the *Struan* and the *Tamar E. Marshall*, flown from separate signal halyards during the ceremony, were sent home with the descendants of each respective vessel's builder.

Chapter XX

The *Struan*'s Windlass

One of the items on display at Camp Meriwether, near the *Struan* anchors and anchor chains, is the main body of a ship's windlass, about which, until very recently, little information was available. Some old timers—even persons with a seafaring background—believed that it was of too-recent design to have been part of the *Struan*. Others, mindful that it had been recovered from the *Struan* wreck site at a very low tide during the 1950s (with cooperation from some military vehicles on a training project), insisted that it had been aboard the *Struan* when the ship came ashore in 1890.

With some help from the computer internet and specific suggestions by Lars Bruzelius of the University of Upsala in Sweden, we have located patent drawings and a sales brochure type of illustration exactly describing the windlass at Camp Meriwether. Positive identification as part of the *Struan* was made possible by an article in a Saint John, N.B. newspaper some four months after launching of the *Struan*, when, commenting on the completion of another ship by the *Struan*'s builder, it was indicated that shipbuilder John Fraser was equipping each of his vessels with an Emerson's Patent Windlass.[1]

Inventor James Emerson assigned the entire interest in his invention of the windlass (for which U.S. Patent No. 12718 had been granted April 17, 1855) to William L. Gilmore of Worcester, Massachusetts, who believed Emerson's letters patent to be "inoperative and invalid by reason of a defective and insufficient description or specification and claim." Accordingly, Gilmore applied for a reissue based on a more detailed description and explanation that he submitted, and was granted reissue status (No. 1020) on July 1, 1860. (We

[1] The Saint John, New Brunswick *Globe*, August 9, 1877, in an article headed "Ship Launches."

found Gilmore's description of the operation of the device more lengthy, but just as confusing as Emerson's had been).

Emerson's windlass (with reference to the diagram from the *Boston Shipping List* of November 22, 1856) consisted of a large horizontal steel shaft turning in heavy bearings installed in the two main supporting bolsters (labeled "K"). Mounted on that shaft were two permanently attached (keyed to the shaft) ring gears, one having about 36 teeth and the other about 18 teeth, each of which was furnished with its own fluted pulley sprocket, called a wildcat, which controlled one of the anchor chains.

It was possible by use of a slip-ring device to disconnect either or both of these chain-bearing sprockets from their companion ring gears, thereby allowing the sprocket carrying the anchor chain to "run free" (turn on the shaft) while subject to control only by the brake bands (labeled "M") which could be tightened by exerting pressure on a lever handle not illustrated in the sketch. The horizontal shaft extended at each end through its bearing in the supporting bolster to accommodate a spool or drum (also permanently keyed to the horizontal shaft) by which the windlass could be used as a winch to handle heavy hoisting or pulling loads, using two or three turns of heavy manila rope around the drum to give friction to the hoisting or pulling line.

The *Struan* Capstan
— *1998 photo by the author*

The capstan, as illustrated in the advertising sketch, was installed on the weather deck of a vessel directly above the windlass, in the *Struan*'s case on the forecastle (fo'c's'le) head—topped with two horizontally mounted wheel-like appendages, one above the other, each having sockets for insertion of the stout oak handles or capstan bars utilized by sailors to operate the windlass.

A comparison of the 1856 inventor's sketch of the Emerson Ship Windlass with the photograph of the actual *Struan* capstan (recovered from the wreck site in the 1950s) reveals that sometime in the intervening years between invention of the device and its actual installation on the *Struan* in 1876-77, the design had been simplified to eliminate the double top, and that only one set of four capstan bar holes was available in the model carried by the *Struan*.

It is also interesting to note that although the earlier sketch did not show, nor did the patent application reveal, there was incorporated in the *Struan* installation a geared perimeter section at the base of the capstan, upon which a ratchet device had been mounted to make a fail-safe or dead-man feature that would automatically hold whatever load had been taken in if sailors had momentarily stopped rotating the capstan to rest or await further orders.

Operation of Emerson's Invention

Essentially, operation of Emerson's windlass required four sailors to walk around the capstan, each pushing against his own stout oak handle or capstan bar (inserted in one of the sockets in the capstan head) to rotate that device in one direction or the other. A three and one-half inch diameter vertical steel shaft permanently fixed to the capstan head transferred that movement, if the upper section of the capstan-head were being utilized, to the small pinion gear "C" in the windlass itself on the deck below, which action caused the large ring gear (at right center in the 1856 illustration) to rotate slowly around the heavy horizontal main shaft, carrying with it the temporarily attached chain-supporting fluted pulley-sprocket, thereby bringing in the *starboard* anchor chain. When the *bottom* half of the capstan head was rotated in the opposite direction, another pinion gear "D", mounted lower on the vertical shaft, acted similarly against the *small* ring gear (at left center in the illustration), bringing in the *port* anchor chain.

The two pinion gears mounted on the vertical shaft, "C" and "D" as noted above, were each equipped with a simple ratchet-clutch-pawl device that permitted one or the other to slip freely on the vertical shaft when it was turned in one direction, while turning the capstan head in the other direction caused one or the other of the pinion gears to engage either the large or small ring gears. No matter which way the capstan heads were rotated, one, *but never both* of the ring gears would thereby be activated.

From an inspection of the *Struan*'s windlass after its 60-year immersion in salt water, and careful study of the 1856 illustration, it appears that the small pinion gear "C" (with about 12 teeth) would rotate the large ring gear (with about 36 teeth) approximately one-third of a revolution for each 360 degree movement of the capstan head. Because the fluted pulley sprocket (sometimes called a wildcat) by which the starboard anchor chain could be controlled was substantially smaller than the large ring gear, and knowing from earlier observation[2] of the chain recovered in 1890-91 just after the stranding that each link originally averaged slightly more than 9-1/2 inches in length, it can be estimated that about 54 inches (4 feet, 6 inches, or 4.5 feet) of chain came in with each three complete revolutions of the capstan when the large ring gear was being utilized. To raise the starboard anchor when 30 fathoms (180 feet) of chain was out, then, would have required the sailors manning the capstan to complete 120 full revolutions (walk arounds).

The small ring gear controlling the port anchor had perhaps 18 teeth, meaning that it would make two-thirds of a revolution for each complete turn of the capstan head, and would have brought in the same amount of chain with only 1-1/2 revolutions of the capstan—essentially at twice the speed but one-half the power available if the large ring gear were involved. The obvious conclusion is that this higher speed but lower-powered section of the windlass would normally have been reserved for handling lighter loads. By disconnecting the two pulley sprockets (wildcats) from their adjacent ring gears and utilizing a series of snatch blocks and a heavy manila line, the windlass could have been used to bring down a damaged spar or send up a replacement, etc.

As the incoming chain came off one or the other of the pulley-sprockets, it was automatically directed through a hole in the deck, and fell of its own weight into the chain locker in the vessel's hold below the windlass. Each anchor chain, sometimes called a cable although made up of a combination of individual links of iron or steel chain, had its own chain locker to avoid the possibility of one chain and its links being entangled with other chains as they were taken in and dropped by gravity from the deck on which the windlass was installed. Such a complication in chain storage could easily have resulted in a delay or even a complete inability to pay out chain when an anchor was unexpectedly needed in an emergency.

To let the anchor out, a clutch-like arrangement could be activated to disconnect the chain-bearing pulley sprocket from its adjacent ring gear, allowing the anchor to "run free" of its own weight toward

---

[2] This author first saw the *Struan* anchor chain during the summer of 1931. At that time, some vestiges of black paint could be detected on most of the links, which were but little reduced from their original wire size of 1-5/8th inch diameter iron. Whether that black paint remained from one of the many coats applied by crews under the four captains known to have commanded the vessel or had been put in place by early-day scouts at Camp Meriwether to "dress up" the anchor and chain display around the flagpole, cannot be known for certain.

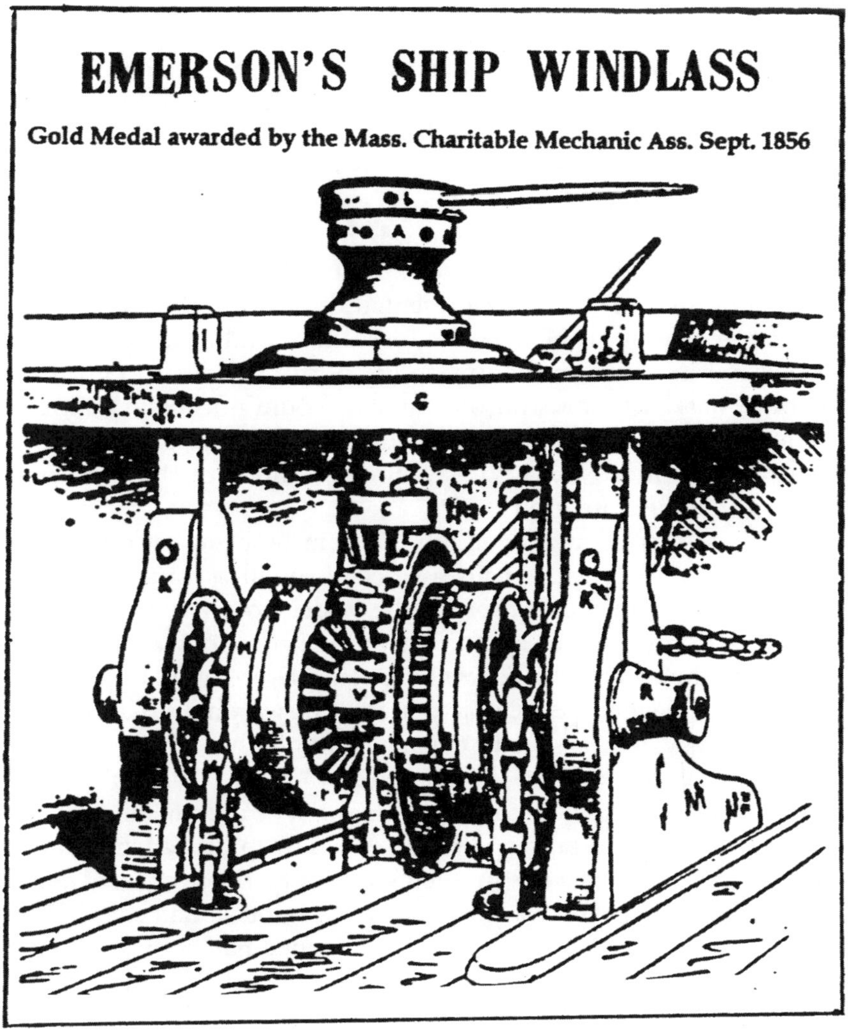

Illustration above: Illustration from an advertisement published in the "Boston Shipping List" November 22, 1856, showing details of Emerson's Patent Ship Windlass. A later version of similar design was installed in the *Struan* during construction of the vessel in 1876-77.

Photograph opposite: Ship's windlass recovered from the surf during an extremely low tide at Camp Meriwether in the 1950s, at the location where the *Struan*'s anchors and anchor chains were salvaged by homesteaders in 1890-91.
— 1990 photograph by Joan Crichton

the bottom through its hawsepipe in the bow of the ship, controlled only by the brakes on the windlass.[3]

In this connection, it was very important that a shipmaster know the approximate depth of the water under his vessel before making a decision to anchor. This was done in the *Struan*'s time by sounding with a hand lead on a light line. It was also necessary for the captain and all his mates to be absolutely certain about the length of the chain carried for each anchor. Although the chain could usually be allowed to run free at whatever speed the weight of anchor and chain made applicable in comparatively shallow locations (under about ninety feet or 15 fathoms), deeper water presented difficulties in controlling the run-out. A serious problem could arise if the combined weight of the anchor and its chain on the way to the bottom became so great that the brakes and set-works on the windlass could not retain control of the load. For that reason, shipmasters were advised to pay out the anchor chains more slowly, always under control of the windlass in deeper anchorage situations.

To avoid loss of an anchor and its chain by being let out too far, the "bitter end" of each chain was led through a massive ring bolt firmly anchored in the keelson of the vessel, to a heavily reinforced pad-eye 10 or 12 feet away, so that the impact shock occasioned by coming to the end of a chain would be transferred away from a single location and divided between two areas.[4] In wartime situations or in any instance where a speedy departure might have been desirable, some shipmasters attached the "bitter end" of their anchor cables to the vessel with an easily released pelican hook or even manila lashings. Under those circumstances, a vessel could "slip" its anchor and be ready for departure in a few moments without a thirty or forty minute delay while the anchor was noisily and laboriously hauled in. When rope lashings were the choice, a well-sharpened hatchet was left nearby for instant use if the command came to "let go the bitter end."

To avoid fouling the hawsepipes, windlass, and chain locker with sand, mud or debris from the bottom, it was customary when taking aboard an anchor, in the *Struan*'s time, to have a sailor direct a stream of seawater from a hand-operated high-pressure pump onto the anchor chain as it entered the hawsepipe. Today's sailors use high-pressure fire hoses for the same purpose. On some well-kept vessels in the *Struan*'s period, when weather and scheduling constraints permitted, anchor chains were brought aboard a little at a time, thoroughly

---

[3] Edward A. Turpin and William A. MacEwen, *Merchant Marine Officers Handbook* (Centerville, Maryland: Cornell Maritime Press 1965), p. 11-7.

[4] Turpin and MacEwen, p. 11-10.

cleaned, allowed to dry, hammer tested for defects and painted in place before being dropped into the chain lockers. In today's merchant marine, with ships carrying minimum crews barely able to operate the vessel under normal conditions, maintenance work of this nature is usually put off to be done at shipyards during annual overhauls.

Prior to about 1924, wrought iron chain was used for making up ships' anchor cables, and the studs (cross members connecting the two sides of each link) were individually welded into place.[5] Currently, cast steel chain is produced with the studs cast integrally with the link itself.

Aboard today's U.S. merchant vessels, anchor chain is made up of 15-fathom (90 foot) sections called shots. Any number of shots can be joined together with detachable links to produce the required chain length. Whether or not this practice was in common use internationally 110 years ago has not been established with any degree of certainty.

Required weights for anchors, sizes for anchor chains, and lengths of anchor cable to be carried aboard vessels of various tonnages are today all spelled out in "Rules for Building and Classing Steel Vessels," published by the American Bureau of Shipping, sometimes known in earlier years as "American Lloyds." The 1917 edition of that volume (earlier editions from 1867 have escaped our search) specifies that a sailing vessel of the *Struan*'s "equipment weight" (which very closely approximates the gross tonnage of a vessel) must carry two bower anchors each weighing 3,465 pounds, a spare bower anchor weighing 2,940 pounds, a lighter stream anchor weighing 1,435 pounds, and at least 240 fathoms (1,440 feet) of 1 5/8ths inch (wire size) studded chain. Presumably the chain would be divided to assign half of it to each of the heavy bow anchors.

Each detachable link between shots is painted red for easy identification. One link on each side of the first detachable red link (15 fathoms from the anchor) is painted white, and the stud on those two white links shows one turn of wire. At the end of the second shot (thirty fathoms from the anchor) there are *two* white links on each side of the detachable red connecting link and *two* turns of wire on the second stud on each side of the connecting link. This marking system is continued throughout the entire length of the anchor cable by

[5] During the summer of 1990, while cutting apart a short section of the *Struan* anchor chain to facilitate presentation of an individual link to each Norwegian and Canadian guest at the *Struan* Centennial Celebration, it was discovered that several stud link welds were defective. In at least one instance, the hand-welded stud fell free of the link, leaving no visible indication either on the link or the stud itself that there had ever been a permanent attachment.

adding a white link on either side of the red connecting link and making an additional turn of wire at each 15-fathom connection. Through use of these standardized markings on all vessels, the officer on the bow can observe the chain being payed out and have a reasonable estimate at any moment of how much cable is in the water. Most mariners accept the generally recognized rule of thumb that dictates riding at anchor to a cable length from five to seven times water depth under normal conditions. More chain was normally let out when preparing to ride out a storm in an unprotected anchorage since mariners believe that the more cable is out, the more easily a vessel rides to a gale and that there is less danger of dragging the anchor.[6]

Once the anchor was on the bottom and the appropriate length of cable run out, a clamping device, securely fastened to a reinforced deck section forward of the windlass, was used to grip a link of the anchor chain to take the strain off the windlass itself for whatever period the vessel remained at anchor.[7] These clamps are commonly known as chain stoppers.

A swivel link, usually mounted in the anchor cable about five chain links from the anchor shackle itself,[8] prevented kinks or twists in the cable as one or the other of the anchor flukes bit into the sea bottom. At congested anchorages or at any location where an anchored vessel could not safely be permitted to swing or turn freely with the wind or tide, it was (and is) sometimes necessary to put out a stern anchor to hold the ship in position. In most instances, this would have been the stream anchor all vessels were required to carry.

Since few vessels in the *Struan*'s time had a windlass at the stern to handle an anchor chain, any anchor put out from the stern would have been connected to a wire hawser or, if necessary, to a manila towing hawser.

The dictionary definition of a "cable length" in the United States is 720 feet, which works out to eight shots of 15 fathoms (ninety feet) each. In Great Britain, the cable length is 608 feet, although neither of these lengths should be arbitrarily accepted as an indication of the amount of anchor cable available on any particular vessel. Although we have discovered no evidence to indicate the length of chain carried aboard the *Struan*, there are recorded instances of vessels at deeper Puget Sound anchorages riding to 500 or 550 feet of anchor cable. No information has come to our attention regarding a possible British Board of Trade requirement for a specific length of chain to have been

---

6 Austin M. Knight, *Modern Seamanship* (Publisher not identified, 1944) p. 295.

7 Turpin and MacEwen, p. 20-63.

8 Turpin and MacEwen, p. 11-9.

furnished with a new vessel at the time of the *Struan*'s launching.

One of the advantages of having the anchor chain made up of easy-to-disconnect 90-foot sections (shots) is the relative simplicity of repositioning the 15-fathom lengths in their sequence from the anchor itself so as to equalize wear on the various sections. Shots from cable near the "bitter end" (which seldom saw any actual use) should be moved into the most worn location adjacent to the anchor during the annual (or more frequent) overhaul of all the ground tackle. After a close inspection and minor repainting of links and some changes in wire wrappings, the anchor cable could be cleaned, painted, and prepared for dependable further service.

Nineteenth century mariners were fully aware that the design and size of the flukes were more of a determining factor in the holding power of their vessel's anchors than the dead weight of the anchors themselves. So long as sufficient chain (cable) had been let out to permit the chain to pull almost parallel to the bottom, any strain on the cable caused one fluke or the other to dig into whatever the bottom offered—mud, sand, or shingle (gravel and small stones).

One disadvantage of the old-fashioned anchor (with its fixed crosswise stock) was its propensity to cause fouling of the anchor chain, particularly if the anchor was dropped while the vessel was stationary. Extra chain lowered on top of an anchor already on the bottom invited snagging of the chain on the stock or one of the flukes, sometimes resulting not only in failure to hold the vessel in place, but also in time-consuming difficulty in raising the anchor and getting it safely aboard and in position to be used again. For that reason, a shipmaster was advised to have his vessel slowly underway when dropping the anchor. Some cautious captains of square-rigged ships like the *Struan* would back their sails to give sternway to the vessel as the anchor chain ran out, safely away from the anchor itself. Other masters, when anchoring in a river or area with a tidal flow, would head upstream and pay out chain until the anchor touched bottom, and then let the current back the vessel away from the anchor as more chain was released.

From my first inspection of the *Struan* anchor chain in 1931, before seacoast conditions had rusted the links to the finger-sized wire diameter reflected in some of the cable still on display, it is estimated that the chain was originally made up of links about 1 5/8ths inch diameter "wire size," making a fathom (six feet) of chain weigh about 158 pounds.

Although the "old-fashioned anchor" as used by the *Struan* has long been recognized as more efficient and better holding than the modern day patent or stockless anchor (and under normal conditions requires less scope of chain length to hold a vessel safely in place) it is almost unknown in today's merchant fleets because of its difficult and awkward handling characteristics. Where a patent or stockless anchor can be drawn up snugly by the windlass into the hawsepipes ready for almost instantaneous use, the stock on the "old fashioned" anchor required that it be handled very carefully as in broke water when being taken in, and making it fast in any kind of bad weather became both difficult and dangerous.

One practice favored by some captains in the *Struan*'s time, apparently including Captain Skogland of that vessel, involved coupling a heavy weight to the inboard end of the first shot of chain let out, thereby assuring that at least ninety feet of cable would lie on the bottom and provide maximum holding power for the anchor.

An iron ball about 16 inches in diameter, probably used in this manner, was recovered from the surf at the *Struan* wreck site in the 1950s.

> Using the formula $4\pi r^3 \div 3$ for the volume of a sphere, and accepting .2833 pounds as the weight of a cubic inch of iron, it can be calculated that the weight of the iron ball is about 608 pounds.

Use of such a weight on the anchor cable, particularly if connected closer to the vessel than on the first shot, would have been wisely appropriate during the *Struan*'s stay at Iquique, Chile, where ships at anchor there were constantly in danger of tidal surges occasioned by earthquakes at sea. If a considerable length of chain was out when such an event occurred, the attached weight would have acted as a shock absorber, allowing the ship to fall back only a short distance while the force of a giant wave lifted the weight and the lengths of anchor cable off the bottom to approximate a straighter line from anchor to vessel, perhaps eliminating a sudden yank that could have torn out the anchor or snapped the cable.

Epilogue

After transferring the shipwrecked *Struan* crew to the tug *Vigilant* outside San Francisco harbor on January 2nd, 1891, Captain Kenney made good his intention to take the *Tamar E. Marshall* to England without intermediate stops.

He brought the vessel into Sunderland[1] on May 12, 1891. While the cargo of sacked wheat was being unloaded, Kenney and his first officer, C. Fred Leary, lost no time in finding a scrivener[2] to prepare a petition to the Norwegian government asking for reimbursement for seven days lost time "off course" while the *Tamar E. Marshall* sailed in and out along the California coast looking for a vessel to take the *Struan*'s rescued crewmen ashore.

Whether that request for monetary repayment was made on his own initiative or by order of the vessel's owners in Canada has not been ascertained by this writer.[3] In the same document, he also claimed payment for the meals eaten by the *Struan* crew while aboard his vessel.

Captain Andrew Freeman Kenney, though in command of the *Tamar E. Marshall* as an interim or relief captain in place of Captain James H. Parker, was an experienced master mariner in the employ of the Troop fleet, and had been dispatched from Saint John to meet the *Marshall* after it arrived at Port Townsend from the Orient, apparently in response to Captain Parker's telegraphic request for temporary relief from command. Parker resumed command on the next trip.

---

1 A port city on the east coast of England, in the county of Northumberland about six miles south of the mouth of the river Tyne, near Newcastle-upon-Tyne.

2 Before the days of typewriters, a scrivener was a person who, for a fee, prepared documents, letters and deeds, etc., by writing them legibly in longhand with pen and ink.

3 Records of the Troop fleet operations after the Great Fire of 1877, if not thrown out when the firm ceased business, have not come to light or become available for public scrutiny.

Captain Kenney, born at Saint John, N.B. in 1845, completed his examination at Liverpool in 1871, and was issued sea captain's certificate No. C 0657. Little is known of his early maritime experience until he associated himself with the Troop fleet of Saint John. In 1875 he was the first commander of the David Lynch-built 1,769 ton ship *Rock Terrace*, pride of the Troop organization and named for the Troop estate. After the Great Fire of 1877, in spite of having lost two Orange Street homes to the flames, he left for England in command of a Troop vessel almost before the ashes had cooled. He saw service as captain of the *Maple Leaf* in 1880-81, *Josie Troop* in 1882, *Cedar Croft* in 1883-84, *Crown Jewel* in 1885, and *Alpheus Marshall* in 1888. His later commands included the *Abram Young* in 1889, *Tamar E. Marshall* in 1890-91, *Cyprus* in 1892, *Privateer* in 1893, *Mistletoe* in 1894-96 and

---

Photograph opposite: Ezra and Hannah Chamberlain, with their children, grandchildren and daughter-in-law, gathered about 1895 in front of the home they built, largely from timbers and lumber washed ashore after the wreck of the *Struan*. Boards on the sturdy fence they put up to keep deer from their flower and vegetable garden, were resawn in a water-powered sawmill constructed on the premises to make better use of the large supply of heavy timbers available after the stranding.

Ezra's great grandson, Robert Chamberlain, recalls his father telling him that the itinerant photographer who took the picture arrived at the homestead leading a burro laden with photographic supplies, camping gear and provisions, and that he stayed only briefly to pose the family members and take the picture. The original photo provided to the Chamberlains bears on its back the imprint of "L. BRADLEY—ARTIST, DAYTON, ORE".

Levi Bradley, reportedly born about 1868 in Ontario, Canada, was listed in the 1888 R.L. Polk and Company's "Oregon, Washington and Idaho Gazetteer" as having a photography business in Yaquina, Oregon. He later operated briefly from Forest Grove, and settled in Dayton, Oregon in 1890, where he advertised as a traveling artist while carrying on his pursuits as a roving photographer.

Several of the period photographs in the possession of the Dayton, Oregon public library bear either his signature or a written indication that he was the "artist." His specialty was persuading family groups to pose in their yards or on the porches of their homes. Businessmen were pictured in front of their shops or stores. Some spectacular shots of the damage caused by the great flood of 1890 to Dayton business properties, while they cannot be positively ascribed to Bradley, are in his style and are generally credited to him.

It is easily understandable that after he photographed everything of interest in his local community, he needed to roam farther afield to find new customers for his business, as was demonstrated by his 60- or 70-mile trip to the vicinity of the *Struan* wrecksite in 1895.

He moved to Seattle in 1903, where he became a long-time photographer for the Seattle police department, from which position he had retired prior to his death in 1952.

— *Photograph from the Chamberlain family, through the Portland Area Council, Boy Scouts of America (1931).*

Nellie Troop in 1896. After the turn of the century he sailed as master of the *Kate F. Troop* in 1900 and commanded another vessel of the Troop fleet in 1901.

The large number of vessels he commanded and the relatively short times he spent on any one craft combine to give credence to the belief that Captain Kenney served the Troop organization as a staff or reserve commander who could quickly be transferred or assigned wherever the need was greatest.

When interim Captain Kenney's petition reached officials at the office of the Swedish and Norwegian Consulate General in London, their reaction was understandably one of surprise and astonishment.

The Norwegians, with hundreds of sailing vessels and a growing fleet of steamships trading all over the world's oceans, were certainly cognizant of the necessity for protecting their seafaring citizens in distress situations, but they were not anxious to open the doors of their treasury to a plethora of claims for funds every time one of their sailors was provided with a meal aboard a rescue vessel.

In his document, Kenney chose his words with extreme care so as to *imply* but not actually *say*, that he had sent a boat to pick up the *Struan* crew from their derelict vessel. Whether King Oscar II and his advisers, while they were considering payment for the *Tamar E. Marshall*'s "off course" sailing, had at hand Captain Skogland's version of events relating to the rescue cannot be determined using available documentation.

To suggest that Kenney's petition might have evolved into a minor international incident may be overstating the importance of the situation as it existed at that time. To keep the matter under control, however, Norwegian authorities made a politely diplomatic inquiry to the British Board of Trade for suggestions and comments as to whether the request for reimbursement for "off course" lost time was reasonable.

With uncharacteristic promptness, the Board of Trade replied within four days (on June 12, 1891) that such payments for "off course losses" in certain circumstances had been allowed where the situation required that rescue vessels actually abandon their original voyage plans to take rescued seafarers to a place of safety. No mention was made in that correspondence of the claim for meal costs for the rescued mariners.

A bill for 70£ sterling—a little over $340 by exchange rates in effect at the time—for "deviation of ship's voyage" and an additional 25£ 13 shillings for meals provided to the shipwrecked mariners then went forward through channels with Kenney's recitation of the circumstances as prepared by the scrivener he had employed.

It is interesting to note in this connection that the *Struan*'s deck boy, Anton Th. Madsen, in his later recollection of the rescue and his

Bermuda Royal Gazette

Hamilton, September 22, 1896.

☞ On Thursday last, the 17th inst, the Norwegian barque "Luiso" called off these islands for the purpose of landing the shipwrecked crew (18 men all told) of the British barque "Tamar E. Marshall", Utley, master, which vessel foundered at sea on the 8th instant. The crew was taken off the barque "Luiso" by the tug "Gladisfen" about 10 miles South of Gibb's Hill Light (the barque proceeding on her voyage) and landed at St. George's and placed in charge of J. H. S. Frith, Esq., Assistant Receiver General and Controller of Customs, etc., at that port. Capt. Utley, of the British barque "Tamar E. Marshall", on a voyage from Savannah bound to Santos, S.A., with a cargo of timber, reports having left Savannah on the 27th Aug. last; had fine weather up to the 7th instant. On that date in latitude 34.20 N., longitude 73.20 W., experienced a heavy hurricane from the N.E. Had bulwarks smashed in, deckload washed away, and the vessel became waterlogged. On the 9th the Norwegian barque "Luiso" coming in sight, made signals of distress to her, which she immediately answered; bore down and hove to and succeeded, with much difficulty (owing to the tremendous sea running at the time), in taking the crew off. The Capt. of the "Luiso" then decided to make for Bermuda for the purpose of landing the crew at that port. Made land on the 17th, when the crew were landed as above reported. Capt. Utley, with his first and second mates and eight men will leave by the SS. "Duart Castle" for St. John, N.B., either to-day or to-morrow. The remaining seven men will leave for New York by the SS. "Orinoco" on her next trip.

Reconstruction of an 1896 news article with corrections made to eliminate originally published errors.

time aboard the *Tamar E. Marshall*, recalled that "we thought we got rather little at each meal—however it was (perhaps) best so."

The Norwegian Minister of the Interior in Royal Council recommended payment only of the "deviation from course" portion of the claim, perhaps because it appeared to him on close examination to be for less than previous recoveries by other claimants against a number of maritime nations under relatively similar circumstances. King Oscar II wrote "Expederes" (Ekspederes) meaning "dispatch it or send it,"—literally translated "pay it"—at the bottom of the paperwork and signed his bold "Oscar" to bring the matter to a close. So far as can be determined, no further reference to the subject of reimbursement for meals served the rescued sailors was ever made by either party.

In retrospect, an examination of Captain Kenney's actions after taking the crew of the *Struan* aboard his vessel might be of interest. Kenney reported in his claim to the Norwegian government that he

Shipowner (Skibsreder) Tollak Johan Hansen Skogland, 1857-1934
Master of the *Struan* at the time of its loss.

— Photograph (circa 1920). Courtesy Rederiforeningens Museum, Haugesund, Norway

From the left, Daryl Fraser, great grandson of *Struan*'s builder John Fraser; the author; and Dr. Clarke Fraser, great grandson of Alpheus Marshall, builder and major owner of the *Tamar E. Marshall*, which rescued the *Struan* crew at sea in December 1890. Picture taken at Digby, Nova Scotia in the summer 1991 just before Dr. Fraser hosted the trio at his Bear River home, which had originally belonged to Captain James H. Parker, long-time skipper of the *Tamar E. Marshall*. Following lunch on the veranda, Dr. Fraser cooperated in a search of the premises for photos, paintings, log books or other records pertaining to the vessel, but nothing was found.

—Photo by L. Hopkins, Bear River, Nova Scotia. Used by permission.

had picked up the *Struan* seafarers on December 16 in latitude 42.46 N, longitude 128.50 W. Captain Skogland of the *Struan*, in his interview with news reporters in San Francisco upon his arrival there, confirmed that they had abandoned the *Struan* in approximate latitude 43°N.

The position of the rescue as plotted by Kenney is some 300 miles off the Southern Oregon coast opposite Port Orford, which is about 375 miles north of San Francisco.

The *Struan* had left Port Townsend November 24, and in spite of terrible weather had progressed down the coast to the point of rescue—or even further south[4]—by the time the sails were all blown away December 3rd. In about ten days, then, the *Struan* had sailed about 1-2/3 times the distance from the point of rescue to San Francisco.

Even if the undamaged *Tamar E. Marshall* made only the same speed as had the *Struan*, it could theoretically have arrived outside the San Francisco harbor within seven days of encountering the *Struan*—a voyage contemplating a straightline distance of some five hundred miles involving a much greater probability of meeting some inbound vessel approaching that port.

Kenney chose instead to sail in and out along the Northern California coastline (where there were not many well-developed deepwater ports and few large vessels using them) in the hope of finding some captain to relieve him of the shipwrecked mariners. Accordingly, the *Struan* crewmen had been aboard the *Tamar E. Marshall* for 17 days by the time the tug *Vigilant* came alongside to take them to San Francisco.

During that time, the *Marshall* sighted only one other vessel, the American bark *George S. Homer*, a converted steamer, whose master summarily refused to take the *Struan* crew on board, though he did agree to sell some provisions to the rescue vessel to alleviate the shortage brought on by the extra 18 men being fed.[5]

After-the-game quarterbacking is always easy to do and hard to excuse. In this case, however, it seems fair to suggest that the seven days "off course" lost time for which his owners were compensated by the Norwegian government were really a result of a conscientious choice Kenney made on his own. If he had set course immediately after the rescue directly for San Francisco, he might have been able to begin his voyage to England ten days sooner.

On September 7, 1896, while in the command of Captain Charles

---

4 During mid-December 1890, newspapers along the Northern California and Southern Oregon coasts had been reporting sightings by various mariners of a dismasted derelict in the area off Point St. George, about twenty miles south of the Oregon-California border. Although some observers thought the derelict to be another vessel, others reported it as the *Struan*. One shipmaster recalled that he had believed the derelict to be a new vessel because the metal covering on the hull had been shiny and bright.

5 Captain Kenney's erroneous description, in his claim against the Norwegian government, of the location at which he encountered the *George S. Homer* (lat. 37.39 N., long. 128.11 E), must have amused the staffers who handled it if any of them read it carefully. The position he reported being in as of December 29, 1890, four days before he transferred the *Struan* mariners to a tug off the Farallones, *is in the Sea of Japan off the coast of North Korea*, some five thousand miles from where he actually was. His meeting with the *George S. Homer* was obviously in 128°11' *West* longitude.
The Farallones are a group of six small islands lying about thirty miles off the coast of mainland California, forming a ten-mile long chain paralleling the coastline. They are now a part of the city of San Francisco.

E. Utley,[6] the Canadian bark *Tamar E. Marshall*, ten days out from Savannah, Georgia for South America with a cargo of lumber, encountered a "heavy" hurricane from the Northeast. The bulwarks were smashed in, the deck cargo washed away, and the hull began to leak so badly that the vessel became completely waterlogged.

The *Tamar E. Marshall* found itself in exactly the same condition and predicament in the Atlantic as it had found the *Struan* almost six years earlier in the Pacific.

Luckily for those on board, a Norwegian bark, the *Louiso*, happened by within about 48 hours and "with much difficulty" rescued all 18 men on board the unfortunate vessel. The Norwegian captain changed course immediately for Bermuda, where the shipwrecked crew was landed about ten days later.

Though such a discovery might have made history more interesting by providing a decidedly ironic or satirical twist, we have found no indication of a request by the Norwegian bark's owners for payment either for "deviation from course" or for meals served the shipwrecked British mariners while aboard the Norwegian rescue vessel.

Of the 18 men who reached San Francisco in January 1891 after abandoning the *Struan* at sea for the safety of the *Tamar E. Marshall*, only five are known to have returned to Norway after the loss of their ship.

Captain Tollak Johan Hansen Skogland went home to his farm at Skare near Haugesund, and within a few years began to accumulate the first of some thirty steamships (and one barque) he would eventually own as he built up his fleet with particular attention to the Hamburg-Buenos Aires route. In 1913 his son Valdemar became a co-owner of the company and the name was changed to "T.H. Skogland & Son." About that time Captain Skogland moved into town to be in closer touch with his various maritime interests. A 1920 photograph looking across the waterway from the southwest shows Skogland's home on the hill overlooking the Haugesund harbor, and a street named for him leading down to the water's edge, where his dock and warehouses,

---

[6] A Nova Scotia shipmaster, born at Yarmouth in 1860, licensed as a captain in 1886. Prior to his service as master of the *Tamar E. Marshall*, he had commanded several other vessels, including the bark *Low Wood*, 1,091 tons, built in the Lynch yard at Portland, Saint John, New Brunswick in 1878 for the Troop fleet. Before that assignment he had been master of another Troop-managed vessel not positively identified because of the near illegibility of old records. Earlier, as mate on the *Cambridge*, he made an almost year-long voyage to South America.

[7] *Det Gamle Haugesund*(Haugesund, Norway: Haugesunds Avis Forlag, 1975), p. 123.

storage yard and coaling facilities were located.[7]

First mate Svend Hoie passed his captain's examination within six months of his return, and pursued that vocation almost continuously until he was lost at sea in 1907 while in command of the bark *Guiana*, which disappeared on a voyage from Mobile to a South American port. At that time the *Guiana*, built at Tusket, Nova Scotia in 1882, had been purchased and was being operated by Haugesund shipowner Johan Thorsen.

Back in Norway after his lengthy hospitalization, third mate John Mandius Hauge I (whose uncle was the well-known Haugesund shipbuilder John Hauge) sailed intermittently for two or three years until the after effects of his *Struan* injuries made sea life difficult for him. In 1882 he established the Viking mineral water factory and began a ship brokering business. He branched out into owning vessels, and by 1899 had a fleet of both sailing craft and steamships (dampskips) that he operated until his death in 1915.

The Hauge family has continued its seafaring tradition for at least four generations. John M. Hauge II was born in Haugesund in 1906. He studied economics at Olso and in 1927 began a seven-year stint working for shipping-related firms in Germany and Italy utilizing his fluency in both those languages as well as Norwegian. In 1934 he opened a small ship brokerage firm which he operated until his death in 1973.

John M. Hauge III was born in Haugesund in 1935. After completing his economic studies at age twenty, he worked for various shipbrokering companies in Haugesund, Liverpool, Paris and Oslo. In 1967 he joined shipowner Hilmar Reksten in Bergen, Norway—one of the leading tanker owners in the world—and for some time has been Reksten's Chartering Manager, and from 1974 a Director and Vice President of the firm.

John M. Hauge IV was born in Oslo in 1962. He finished compulsory school at 16 and attended the Norwegian Government's Sea Aspirant School in Bergen, where he specialized in mechanical studies rather than navigational pursuits. During training he served on some of the world's biggest tankers. In 1984 he was certified as a second marine engineer, and in 1986 as a mechanical engineer from Haugesund Maritime College. For four years he worked for Aker Stord A/S helping to build the Gullfaks C platform and other offshore units now operating in the North Sea.

John M. Hauge V, who, four months before his birth, had accompanied his father John M. IV and his mother Berit to the *Struan* Shipwreck Centennial at Camp Meriwether in August 1990, was born

Four Generations of John Mandius Hauge

I
Third mate aboard the *Struan*
shipbroker/shipowner
1868-1915

II
Shipbroker
1906-1973

III
Ship chartering manager
1935-

IV
Marine/mechanical engineer
1962-

The Fifth Generation

John Mandius Hauge V
Born in Bergen, Norway
December 18, 1990

in Bergen on December 18, 1990.

Anton Th. Madsen, originally carried on the *Struan*'s roster as deck boy, was identified as a seaman by Captain Skogland on the crew list he prepared for Norwegian consul Henry Lund upon arrival at San Francisco aboard the tug *Vigilant*. He continued his seagoing career, served for a time aboard the full-rigger *Noreg*, owned by the Kongshavns of Haugesund, and was last listed as second mate aboard a Norwegian vessel before he retired to a shore-side job of painting houses in Haugesund.

Hans Christian Baardsen, who deserted the *Struan* at Iquique while the coal cargo was being unloaded, turned up in Haugesund a year and a half after other *Struan* veterans arrived there, having worked his way home from South America. In spite of his desertion record, he progressed to mate and was certified as a captain in 1897. In later life he opened a ship chandlery business on the Haugesund waterfront.

The other members of the crew were reported to have "gone out into the countryside" to seek work in the San Francisco area.

Appendix A
Coal Cargoes Carried by the *Struan*

Year	Loading Port	Captain	Discharge Port	Tons Transported	Voyage Duration (Days)	Consignee
1877[1]	Liverpool	Cooper	Bombay	unknown	135	unknown
1879	Cardiff	Cooper	Montevideo	unknown[2]	145	unknown
1880	Liverpool	Cooper	Karachi	unknown	162	unknown
1881	Newcastle, N.S.W.	Cooper	San Francisco	2,177	101	Geo. J. Theobald & Co.
1882	Newport, Wales	Cooper	Panama	2,400	126	Pacific Steam Navigation Co.
1883-84	Newcastle, N.S.W.	Cooper	Panama	2,250	122	Pacific Steam Navigation Co.
1886	Liverpool	Lemon	Rio de Janeiro	2,436	62	Rio de Janeiro Gas Co.
1887	Cardiff, Wales	Scott	Valparaiso	2,472	168	unknown
1888	Cardiff, Wales	Scott	Rio de Janeiro	2,452	45	Norton Megam & Co.
1890	Birkenhead, England	Skogland	Iquique, Chile	2,650[3]	203[4]	unknown

[1] Description of cargo on this voyage cannot be positively verified using available information.

[2] Exact coal tonnage taken aboard for this voyage cannot be positively determined with available data.

[3] Note that after being registered under the Norwegian flag and no longer subject to British load-line regulations, 200 to 300 more tons of coal were loaded by Captain Skogland than had been carried by English-licensed masters governed by rules in effect for British-registered vessels.

[4] Ship returned to Plymouth, England for repairs after storm damage off the coast of France. Total time lost, approximately one month.

Appendix B
Deaths Aboard the *Struan*

British maritime regulations in the *Struan*'s time required that all deaths of crew members be recorded not only in the official ship's logbook, but also on the vessel's "articles," which were essentially the terms of the contract entered into by each crewman as he signed aboard at the beginning of a voyage.

Each line on the "articles" (also known as the crew list) detailed an individual seaman's name, birthplace, age or birth date, time and place of signing aboard, time and place and reason for discharge as well as the amount of money paid the sailor for the time served. Deaths aboard ship also necessitated the filing of death notice form W & E1 with consular representatives or port authorities at the next stop on the voyage.

Although the British government decided years ago that it couldn't continue to provide storage space for all the logbooks it had accumulated, it did retain and attempt to preserve a great majority of the crew lists (articles) by parceling them out to interested universities, museums and similar organizations which agreed to hold them available for study in future years.

Crew lists for all the voyages of the *Struan* under the British flag, both while owned in Canada or England, have been located from a number of these cooperating sources, and all deaths of crew members aboard the vessel (Official Number 72293) are enumerated below.

Voyage begun 11 September 1877 at Liverpool, completed 9 October 1878 at Hamburg:

- Swedish seaman Brent Benson, age 29, died of dysentery near East Point, en route to Rangoon, 3 April 1878.
- Seaman Phillip Gomez died at sea of what the captain recorded as "Cheab's Disease", 30 August 1878.

Voyage begun 27 November 1878 at Hamburg, completed 15 July 1880 at Dunkirk:

- Seaman Edward Anderson washed overboard at sea, drowned, 16 January 1879.

Voyage begun 18 May 1887 at Cardiff, completed 21 October 1889 at Liverpool:

- Seaman Louis Oliver, born in France, age 43, died at sea 10 November 1887 en route to Valparaiso. Captain did not know or did not report cause of death.
- Seaman S.O. Gorgensen, Norwegian, age 26, fell off a scaffold while painting the ship outside the starboard bow and drowned at the vessel's anchorage in New York harbor on 27 August 1889.

No deaths occurred during the *Struan*'s single 14-month voyage under the Norwegian flag. Captain Skogland reported on returning to Plymouth for repair of storm damage that his second mate and three crewmen had sustained storm-related injuries, but each of them remained on the roster. Second mate Thomas Jorgensen went overboard during early stages of the North Pacific storm that resulted in the *Struan*'s loss, but was rescued by Captain Skogland. No details of that incident have survived.

Index